T0297028

'Enriched by many pictures and exercises with solutions, this book provides an accessible and well-written introduction to Lagrangian torus fibrations, which are of interest to a broad audience through their connections with integrable systems and algebraic geometry. This work will be appreciated by students and experts alike.'

— Felix Schenk, *Université de Neuchâtel*

'This is a lucid and engaging introduction to the fascinating world of (almost) toric geometry, in which one can understand the properties of Lagrangian and symplectic submanifolds in four dimensions simply by drawing suitable two-dimensional diagrams. The book has many illustrations and intricate examples.'

— Dusa McDuff, *Barnard College, Columbia University*

LONDON MATHEMATICAL SOCIETY STUDENT TEXTS

Managing Editor: Ian J. Leary,
Mathematical Sciences, University of Southampton, UK

London Mathematical Society Student Texts 105

Lectures on Lagrangian Torus Fibrations

JONNY EVANS

University of Lancaster

CAMBRIDGE
UNIVERSITY PRESS

CAMBRIDGE
UNIVERSITY PRESS

Shaftesbury Road, Cambridge CB2 8EA, United Kingdom

One Liberty Plaza, 20th Floor, New York, NY 10006, USA

477 Williamstown Road, Port Melbourne, VIC 3207, Australia

314–321, 3rd Floor, Plot 3, Splendor Forum, Jasola District Centre, New Delhi – 110025, India

103 Penang Road, #05–06/07, Visioncrest Commercial, Singapore 238467

Cambridge University Press is part of Cambridge University Press & Assessment,
a department of the University of Cambridge.

We share the University's mission to contribute to society through the pursuit of
education, learning and research at the highest international levels of excellence.

www.cambridge.org
Information on this title: www.cambridge.org/9781009372626
DOI: 10.1017/9781009372671

© Jonny Evans 2023

This publication is in copyright. Subject to statutory exception
and to the provisions of relevant collective licensing agreements,
no reproduction of any part may take place without the written
permission of Cambridge University Press & Assessment.

First published 2023

A catalogue record for this publication is available from the British Library.

A Cataloging-in-Publication data record for this book is available from the Library of Congress.

ISBN 978-1-009-37262-6 Hardback
ISBN 978-1-009-37263-3 Paperback

Cambridge University Press & Assessment has no responsibility for the persistence
or accuracy of URLs for external or third-party internet websites referred to in this
publication and does not guarantee that any content on such websites is, or will
remain, accurate or appropriate.

Contents

Contents

Preface

This book is aimed at graduate students and researchers in symplectic geometry. The primary message of the book is that when a symplectic manifold X admits a Lagrangian torus fibration $f\colon X \to B$, the base B inherits an integral affine structure from which we can 'read off' a lot of information about X.

The book is based on a 10-hour lecture series I gave in 2019 for graduate students at the London Taught Course Centre. It also draws on sessions on toric geometry and symplectic reduction which I taught between 2014 and 2017 for the Geometry Topics Course at the London School of Geometry and Number Theory. It is heavily expanded from both of these. It could be used as the basis for a one-semester graduate-level course: the core content is the foundational material in Chapters 1–2, the examples and constructions in Chapters 3–4, and the material on almost toric geometry in Chapters 6–8. The lecturer could then choose whether to include more about Lagrangian submanifolds (Chapter 5 and Appendix H) or about connections to low-dimensional topology or algebraic geometry (Chapters 9–10 and Appendix I).

There are many good books and papers which cover similar ground to this book, including Arnold's book [3] on classical mechanics; Audin's book [5] on torus actions; Duistermaat's paper [26] on action-angle coordinates; Symington's groundbreaking paper [106] on almost toric geometry and her follow-up paper with Leung [66]; Auroux's survey [7] on mirror symmetry, in which almost toric fibrations play a crucial role; Zung's papers [120, 121] on the geometry and topology of Lagrangian fibrations; Vianna's papers [115, 116, 117] on exotic tori and almost toric geometry, and his paper with Cheung [17] on the appearance of mutations in a variety of contexts; Mikhalkin [80] and Matessi's papers [72, 73] on tropical Lagrangian submanifolds. Where there is common ground, I have tried to give a different perspective.

We will not discuss *special* Lagrangian torus fibrations, or much about the connection to *mirror symmetry*. For the reader who is interested in this topic, there are many good places to start, including Kontsevich and Soibelman's influential paper on homological mirror symmetry and torus fibrations [61], Gross's series of papers [45, 46, 47], and much of the early work of Joyce (see, for example, [56]). We will also not get as far as discussing the piecewise-smooth torus fibrations of Castaño-Bernard and Matessi [13, 14], or the far-reaching and highly technical constructions of W.-D. Ruan [87, 88, 89].

Whilst reading, you will see that some lemmas are left as exercises. This is because the proof is either (a) easy, (b) fun, or (c) too much of a distraction from the main narrative.[1] You will find the proofs of these in the sections called 'Solutions to Inline Exercises' at the end of each chapter. There are also extensive appendices: some provide background and make the book more self-contained, some discuss in more detail matters which are mentioned in the main text at a point where a full discussion would distract.

Starting in Chapter 1, I will not assume you already know about symplectic geometry and Lagrangian submanifolds (though it wouldn't hurt). I will assume that you do know the following:

- Differential forms and De Rham cohomology (and occasionally singular homology, though only in passing).
- Lie derivatives, though I have included an appendix (Appendix B) which gives a high-level overview of this topic, including a proof of Cartan's 'magic formulas' for taking Lie derivatives of differential forms.
- Some basic notions from differential topology like submersions, and critical or regular values.
- The fundamental group and the theory of covering spaces.

There will probably be other things that I assume in passing, but these are the most important ingredients. In the remainder of the preface, I will assume familiarity with much more, so that I can put this book in context.

Let X be a symplectic manifold. Roughly speaking, a Lagrangian torus fibration on X is a map $f: X \to B$ with Lagrangian fibres. We usually call the target space B the *base* of the fibration. We will see very early on (Theorem 1.40 and Corollary 1.44) that the regular fibres must be tori, and that we can use the symplectic structure to get a natural local coordinate system on B whose transition maps are integral affine transformations. Moreover, under nice conditions, one can reconstruct X starting from this integral affine manifold B (Theorem 2.26). Since the base has only half as many dimensions as the total space, Lagrangian torus fibrations give us a way of compressing information in

[1] In case (c), you shouldn't feel too bad if you can't figure out the proof for yourself!

a way that helps us to visualise and understand four- or six-dimensional spaces using two- or three-dimensional integral affine geometry.

If we restrict ourselves to *regular* Lagrangian fibrations (with only regular fibres) then we can only study a very restricted class of symplectic manifolds (total spaces of torus bundles over a flat base). For this reason, over the course of the book, we gradually expand the class of critical points that f is allowed to have. In Chapter 3, we introduce *toric critical points*, which naturally appear in the theory of toric varieties. This gives us a wealth of interesting examples like $X = \mathbb{CP}^n$ where the integral affine base is simply a polytope in $\mathbb{R}^n$, and we start to use the integral affine geometry of this polytope to understand the symplectic geometry of X (for example, using *visible Lagrangian submanifolds* in Chapter 5). In Chapter 4, we introduce the *symplectic cut* operation: this widens our class of examples to include things like resolutions of singularities.

In Chapters 6–8, we allow ourselves another type of critical point: the *focus–focus critical point*. This was intensively studied by San Vũ Ngọc [111], who understood the asymptotic behaviour of action coordinates as you approach a focus–focus point; understanding Vũ Ngọc's calculation is the aim of Chapter 6. Margaret Symington [106] developed a general theory of Lagrangian torus fibrations with at worst toric and focus–focus critical points, which she called *almost toric fibrations*. In Chapter 7, we find many examples, including Milnor fibres of cyclic quotient singularities. In Chapter 8, we explain Symington's operations for modifying almost toric fibrations (nodal trades, nodal slides, mutations).

Symington's ideas will allow us to get to our first real highlight: the almost toric fibrations on $\mathbb{CP}^2$ discovered by Vianna in 2013 [115, 116]. In these papers, Vianna discovered infinitely many non-Hamiltonian-isotopic Lagrangian tori in $\mathbb{CP}^2$. These tori are very hard to see in our 'usual' pictures of $\mathbb{CP}^2$, but become very easy to construct and study using almost toric fibrations. We will not develop any of the Floer theory required to distinguish these tori, and refer the interested reader to Auroux's paper [7] for an introduction, to Vianna's papers [115, 116, 117] for details, and to Pascaleff and Tonkonog [85] for later developments. Instead, we content ourselves with the construction of the tori; in general, the methods developed in this book are useful for constructing and visualising, but not so useful for proving constraints.

In Chapter 9, we explain some of the most useful surgery constructions that behave well with respect to almost toric fibrations: non-toric blow-up, and rational blow-up/blow-down:

- If you blow up a toric variety at a toric fixed point then the result is again toric, and the moment polytope is obtained from the original moment polytope by truncating at the vertex corresponding to the fixed point (see Example 4.23). Non-toric blow-up allows us to blow up a point in the toric boundary which is not a toric fixed point and obtain an almost toric fibration on the result. This operation was discovered by Zung [121] and further elaborated by Symington [106].

• Rational blow-up/blow-down is a family of operations which allow us to replace a *chain* of symplectically embedded spheres with a symplectically embedded rational homology ball. The simplest example replaces a single sphere of self-intersection −4 with an open neighbourhood of the zero-section in $T^*\mathbb{RP}^2$. This has proved useful in low-dimensional topology for constructing *small exotic 4-manifolds*.

We will use both non-toric blow-up and rational blow-down to understand Lisca's classification of symplectic fillings of lens spaces. Again, we will give an almost toric construction of all of Lisca's fillings but shy away from proving the classification, as this would require nontrivial input from pseudoholomorphic curve theory.

Finally, in Chapter 10, we will study integral affine cones and see that these correspond to symplectic manifolds with singularities modelled on elliptic and cusp singularities. This will allow us to understand the minimal resolutions of cusp singularities and provide us with an almost toric fibration on a K3 surface. The pictures from this chapter will aid the reader who is interested in reading Engel's beautiful paper [31] on the Looijenga cusp conjecture.

Appendices A–E provide some background material on symplectic linear algebra, complex projective geometry, cotangent bundles, and Moser isotopy, in an effort to make the book more self-contained. Appendix F gives a construction of a toric variety associated to a convex polytope with vertices at integer lattice points, as a more algebro-geometric alternative to the construction using symplectic cuts from Chapter 4. Appendix G discusses the contact geometry and Reeb dynamics of hypersurfaces which are fibred with respect to a Lagrangian torus fibration. Appendix H gives a brief exposition of Mikhalkin's theory of tropical Lagrangian submanifolds. Appendix I explains some of the integral affine geometry behind the Diophantine Markov equation, which underlies Vianna's constructions of almost toric fibrations on $\mathbb{CP}^2$.

My goal in writing this book is to provide you with the tools necessary for you to make your own investigations, to probe hitherto unexplored regions of our most cherished and familiar symplectic manifolds, and to bring back and show me the new things that you find. Appendix J, the final chapter of the book, gives a few open problems as inspiration.

Acknowledgements

The aforementioned papers by Denis Auroux, Margaret Symington, and Renato Vianna have been enormously influential on my thinking and geometric intuition, and this book has grown out of my attempts to spread the appreciation of these papers in the wider geometry community. I have also shared many formative conversations and correspondence on this topic with people including: Denis Auroux, Daniel Cavey, Georgios Dimitroglou Rizell, Paul Hacking, Ailsa Keating, Jarek Kędra, Momchil Konstantinov, Yankı Lekili, Diego Matessi, Mirko Mauri, Emily Maw, Mark McLean, Jie Min, Martin Schwingenheuer, Daniele Sepe, Ivan Smith, Jack Smith, Tobias Sodoge, Dmitry Tonkonog, Giancarlo Urzúa, Renato Vianna, and Chris Wendl. Thanks also to Matt Buck, Yankı Lekili, Patrick Ramsey, and the anonymous referees for careful reading and corrections; to Leo Digiosia for spotting a gap in an earlier attempted proof of Theorem 6.7, and several more typos.

I would like to thank the 2014, 2015, 2016, and 2017 cohorts of graduate students at the London School of Geometry and Number Theory, for their insightful comments and questions during my topics sessions on symplectic reduction and/or toric varieties from which the early parts of these notes developed. Thanks also to the audience for my 2019 lectures at the London Taught Course Centre, whose patience and endurance was tested by listening to this material in two five-hour blocks, and whose unflagging cheerful engagement and repartee helped me to improve these notes immeasurably.

Notation

One point of confusion will be the fact that I often take vectors to be row vectors and matrices to 'act' from the right. Apart from the typographical convenience of writing row vectors versus column vectors, this is because my vectors are usually momenta and hence naturally transform as covectors. To remind the reader when I am doing this, I use the convention

$$\begin{pmatrix} a & b \\ c & d \end{pmatrix}$$

to emphasise that a matrix will be acting from the right.

Jonny Evans
Lancaster, 2021

1

The Arnold–Liouville Theorem

1.1 Hamilton's Equations in 2D

Let (p, q) be coordinates on $\mathbb{R}^2$ and $H(p, q)$ be a smooth function. A smooth path $(p(t), q(t))$ is said to satisfy Hamilton's equations for the Hamiltonian H if[1]

$$\dot{p} = -\frac{\partial H}{\partial q}, \qquad \dot{q} = \frac{\partial H}{\partial p}. \tag{1.1}$$

This can be used to describe the classical motion of a particle moving on a one-dimensional line. We think of $q(t)$ as the position of the particle on the line at time t, $p(t)$ as its momentum, and H as its energy. For example, if $H(p, q) = \frac{p^2}{2m}$ (the usual expression for kinetic energy of a particle with mass m) then Hamilton's equations become

$$\dot{p} = 0, \qquad \dot{q} = p/m,$$

which are the statements that (a) there is no force acting and (b) momentum is mass times velocity. You can add in external (conservative) forces by adding potential energy terms to H. Observe that

$$\dot{H} = \frac{\partial H}{\partial p}\dot{p} + \frac{\partial H}{\partial q}\dot{q} = \dot{q}\dot{p} - \dot{p}\dot{q} = 0,$$

so energy is conserved.

From a purely mathematical point of view, Equation (1.1) is a machine for turning the Hamiltonian[2] function $H(p, q)$ into a one-parameter family of maps $\phi_t^H : \mathbb{R}^2 \to \mathbb{R}^2$ called the associated *Hamiltonian flow*. The flow is defined as

---

[1] A dot over a variable stands for differentiation with respect to time, e.g. $\dot{p} = \frac{dp}{dt}$.

[2] Any function can be used as a Hamiltonian, not only ones with physical relevance. The adjective *Hamiltonian* is just here to indicate the way we're using the function H, not that there is anything special about H.

follows:

$$\phi_T^H(p_0, q_0) = (p(T), q(T)),$$

where $(p(t), q(t))$ is the solution to the differential equation (1.1) with $p(0) = p_0$ and $q(0) = q_0$. Conservation of H means that the flow satisfies $H(\phi_t^H(p, q)) = H(p, q)$.

Remark 1.1   In conclusion, given a function H we get a flow ϕ_t^H conserving H. This is a simple instance of *Noether's theorem*. See Section D.3 for a full discussion.

Example 1.2   If $H_1 = \frac{1}{2}(p^2 + q^2)$ then $\dot{p} = -q$, $\dot{q} = p$, so

$$\begin{pmatrix} p(t) \\ q(t) \end{pmatrix} = \begin{pmatrix} \cos t & -\sin t \\ \sin t & \cos t \end{pmatrix} \begin{pmatrix} p(0) \\ q(0) \end{pmatrix}.$$

This corresponds to a rotation of the plane with constant angular speed. Conservation of H_1 means that points stay a fixed distance from the origin.

Example 1.3   If $H_2 = \sqrt{p^2 + q^2}$ then $\dot{p} = -q/H_2$, $\dot{q} = p/H_2$. Since $\dot{H}_2 = 0$, we can treat H_2 as a constant, so the solution is

$$\begin{pmatrix} p(t) \\ q(t) \end{pmatrix} = \begin{pmatrix} \cos(t/H_2) & -\sin(t/H_2) \\ \sin(t/H_2) & \cos(t/H_2) \end{pmatrix} \begin{pmatrix} p(0) \\ q(0) \end{pmatrix}.$$

This flow has the same orbits (circles of radius H_2), but now the orbit at radius H_2 has period $2\pi H_2$.

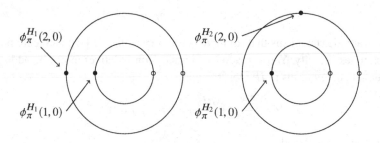

Figure 1.1 Snapshots at $t = \pi$ of the Hamiltonian systems in Examples 1.2 (left) and 1.3 (right) showing the orbits and positions of $\phi_\pi^H(1, 0)$ and $\phi_\pi^H(2, 0)$.

Theorem 1.4   *If all level sets of H are closed (circular) orbits then there exists a diffeomorphism $\alpha \colon \mathbb{R} \to \mathbb{R}$ such that, for the Hamiltonian $\alpha \circ H$, all orbits have period 2π.*

Proof By the chain rule, Hamilton's equations for $\alpha \circ H$ are

$$\dot{p} = -\frac{\partial(\alpha \circ H)}{\partial q} = -\alpha'(H)\frac{\partial H}{\partial q} \quad \text{and} \quad \dot{q} = \frac{\partial(\alpha \circ H)}{\partial p} = \alpha'(H)\frac{\partial H}{\partial p},$$

so the effect of postcomposing H with α is to rescale $(\dot{p}, \dot{q})$ by $\alpha'(H)$. Since H is conserved along orbits, $\alpha'(H)$ is constant along orbits. This means that the orbits of the Hamiltonian flow for $\alpha \circ H$ are just the orbits of the flow for H, traversed at $\alpha'(H)$ times the speed. Let $\Omega := H^{-1}(b)$ be one of the orbits. If the period of the orbit Ω of the flow ϕ_t^H is $T(b)$ then its period under the flow $\phi_t^{\alpha \circ H}$ is $T(b)/\alpha'(b)$. To ensure that all periods are 2π, we should therefore use $\alpha(b) = \frac{1}{2\pi}\int_0^b T(c)\,dc$. $\qquad\square$

Example 1.5 Let us revisit Example 1.3. The period of the orbit $H_2^{-1}(b)$ is $2\pi b$, so the proof of Theorem 1.4 gives us $\alpha(b) = \frac{1}{2\pi}\int_0^b 2\pi c\,dc = b^2/2$. This tells us that to give all orbits the same period, we should use the Hamiltonian $\frac{1}{2}H_2^2$, which is precisely the Hamiltonian H_1 from Example 1.2.

Periods are usually hard to find explicitly; for example, to calculate the period of a simple pendulum in terms of its length, initial displacement, and the gravitational constant, you need to use elliptic functions (see, for example, [42, Chapter 1] or [119, §44]). Similarly, the map α is difficult to write down explicitly in examples. The following theorem gives a useful formula.

Theorem 1.6 *In a 1-parameter family of closed orbits Ω_b, $b \in \mathbb{R}$, of a Hamiltonian system, the period of Ω_b is $\frac{d}{db}\int_{\Omega_b} p\,dq$.*

Proof Assume, for simplicity,[3] that we have coordinates (p, q), with $q \in \mathbb{R}/2\pi\mathbb{Z}$, such that the orbits have the form $\Omega_b := \{(p_b(q), q) : q \in \mathbb{R}/2\pi\mathbb{Z}\}$ for some functions p_b.

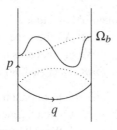

---

[3] One can always find coordinates (p, q) in which the orbits have this form.

Then

$$T(b) = \int_0^{2\pi} \frac{dt}{dq}\,dq = \int_0^{2\pi} \frac{dq}{\dot{q}}$$

$$= \int_0^{2\pi} \frac{dq}{\partial H/\partial p_b} = \int_0^{2\pi} \frac{\partial p_b}{\partial H}\,dq = \frac{d}{db}\int_0^{2\pi} p\,dq. \qquad \square$$

Remark 1.7 This means that $\alpha(b) = \frac{1}{2\pi}\int_{\Omega_b} p\,dq$ is another way of writing the function we found in Theorem 1.4. Note that

$$\alpha(b_1) - \alpha(b_0) = \frac{1}{2\pi}\int_{\Omega_{b_1} - \Omega_{b_0}} p\,dq = \frac{1}{2\pi}\int_C dp \wedge dq,$$

by Stokes's theorem, where C is the cylinder of orbits $\bigcup_{b \in [b_0, b_1]} \Omega_b$. Therefore, if we choose $\alpha(b_0) = 0$, the function $\alpha(b)$ is just the $dp \wedge dq$-*area* of the cylinder connecting Ω_b to Ω_{b_0}.

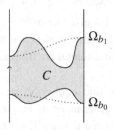

Our goal in this first lecture is to generalise these observations to Hamiltonian systems in higher dimensions. It will be convenient to introduce the language of symplectic geometry.

1.2 Symplectic Geometry

This section uses Lie derivatives, Lie brackets, and the magic formulas that relate these to exterior derivative and interior product; we refer to Appendix B for a quick overview of these concepts and a proof of the magic formulas.

Definition 1.8 Let X be a manifold and ω a 2-form. Let vect(X) denote the space of vector fields on X and $\Omega^1(X)$ the space of 1-forms. Define a map vect(X) $\to \Omega^1(X)$ by $V \mapsto \iota_V \omega$. We say that ω is *nondegenerate* if this map is an isomorphism. A *symplectic form* is a closed, nondegenerate 2-form.

Definition 1.9 Let ω be a symplectic form on a manifold X. Suppose we are given a smooth function $H: X \to \mathbb{R}$. By nondegeneracy of ω, there is a unique vector field V_H such that $\iota_{V_H}\omega = -dH$. We call vector fields arising in this way *Hamiltonian vector fields*. The flow ϕ_t^H along V_H is called a *Hamiltonian flow*.

Example 1.10 Let $\omega = dp \wedge dq$ on $X = \mathbb{R}^2$ and pick a Hamiltonian function $H(p, q)$. Recall that the Hamiltonian flow is defined by $(p(t), q(t)) = \phi_t^H(p(0), q(0))$ and the Hamiltonian vector field is $V_H = (\dot{p}, \dot{q})$. Using the explicit formula for ω, we have $\iota_{V_H} \omega = \dot{p} \, dq - \dot{q} \, dp$. By definition, $\iota_{V_H} \omega = -dH = -\frac{\partial H}{\partial p} dp - \frac{\partial H}{\partial q} dq$. Comparing components, we recover Hamilton's equations:

$$\dot{p} = -\frac{\partial H}{\partial q}, \qquad \dot{q} = \frac{\partial H}{\partial p}.$$

Lemma 1.11 *A Hamiltonian flow ϕ_t^H satisfies*

$$(\phi_t^H)^* \omega = \omega \quad and \quad (\phi_t^H)^* H = H.$$

Proof We have

$$\frac{d}{dt}((\phi_t^H)^* \omega) = (\phi_t^H)^* \mathcal{L}_{V_H} \omega \quad and \quad \frac{d}{dt}((\phi_t^H)^* H) = (\phi_t^H)^* \mathcal{L}_{V_H} H,$$

so it suffices to show that the Lie derivatives $\mathcal{L}_{V_H} \omega$ and $\mathcal{L}_{V_H} H$ vanish. For this, we use Cartan's formula (Equation (B.2)) $\mathcal{L}_V \eta = \iota_V d\eta + d\iota_V \eta$ for the Lie derivative of a differential form η along a vector field V.

We have

$$\mathcal{L}_{V_H} \omega = d\iota_{V_H} \omega + \iota_{V_H} d\omega.$$

Since $d\omega = 0$ the second term vanishes. Since $\iota_{V_H} \omega = -dH$, we get

$$\mathcal{L}_{V_H} \omega = -ddH = 0.$$

Finally, we have $\mathcal{L}_{V_H} H = \iota_{V_H} dH = -\omega(V_H, V_H) = 0$, as ω is antisymmetric.

$\square$

Remark 1.12 Note that if H is also allowed to depend[4] explicitly on t then the previous argument for conservation of energy ($(\phi_t^H)^* H = H$) breaks down; an extra dH_t/dt term appears in $d((\phi_t^{H_t})^* H_t)/dt$. Nonetheless, the flow preserves the symplectic form. For example, consider the Hamiltonian $H_t = t$. We have $\phi_t^{H_t}(x) = x$ for all t, which certainly preserves the symplectic form, but energy changes over time.

Lemma 1.13 *The Lie bracket of two Hamiltonian vector fields V_F and V_G is the Hamiltonian vector field $V_{\{F,G\}}$, where $\{F, G\} = \omega(V_F, V_G)$.*

---

[4] We call a Hamiltonian *autonomous* it does not depend on t and *non-autonomous* otherwise. You should imagine that if H is autonomous then the system is just getting on by itself, whereas if H depends on t then there is some external input changing the system.

Proof By Equation (B.3) in Appendix B, we have $\iota_{[V_F,V_G]}\omega = [\mathcal{L}_{V_F}, \iota_{V_G}]\omega$. Since V_F is Hamiltonian, $\mathcal{L}_{V_F}\omega = 0$. Therefore

$$\iota_{[V_F,V_G]}\omega = \mathcal{L}_{V_F}\iota_{V_G}\omega = d\iota_{V_F}\iota_{V_G}\omega + \iota_{V_F}d\iota_{V_G}\omega.$$

Since $d\iota_{V_G}\omega = -ddG = 0$, we get $\iota_{[V_F,V_G]}\omega = d\iota_{V_F}\iota_{V_G}\omega$. Since $\iota_{V_F}\iota_{V_G}\omega = -\omega(V_F,V_G)$ this tells us that $[V_F,V_G] = V_{\omega(V_F,V_G)}$ as required. $\square$

Definition 1.14 The quantity $\{F,G\} = \omega(V_F,V_G)$ is called the *Poisson bracket* of F and G. We say that F and G *Poisson commute* if $\{F,G\} = 0$.

Remark 1.15 (Exercise 1.45) Recall that the flows along two vector fields commute if and only if the Lie bracket of the vector fields vanishes. Lemma 1.13 shows that two Hamiltonian flows ϕ_t^F and ϕ_t^G commute if and only if the Poisson bracket $\{F,G\}$ is locally constant.

Lemma 1.16 (Exercise 1.46) *Let F and G be smooth functions. Define $F_t(x) := F(\phi_t^G(x))$. Then $\frac{dF_t}{dt} = \{G,F_t\}$.*

Remark 1.17 Lemma 1.16 should look familiar to readers who know some quantum mechanics; it is the classical counterpart of Heisenberg's equation of motion for a quantum observable $\hat{F}$ evolving under the quantum Hamiltonian $\hat{G}$.

1.3 Integrable Hamiltonian Systems

Definition 1.18 (Hamiltonian $\mathbb{R}^n$-actions) Suppose we have a symplectic manifold (X,ω) and a map

$$\boldsymbol{H} = (H_1,\ldots,H_n)\colon X \to \mathbb{R}^n$$

for which the components $H_1,\ldots,H_n$ satisfy $\{H_i,H_j\} = 0$ for all pairs i,j. In what follows, we will assume that the vector fields V_{H_i} can be integrated for all time, so that the flows $\phi_t^{H_i}$ are defined for all $t \in \mathbb{R}$. By Remark 1.15, the flows $\phi_{t_1}^{H_1},\ldots,\phi_{t_n}^{H_n}$ commute with one another and hence define an action of the group $\mathbb{R}^n$ on X. We call this a *Hamiltonian $\mathbb{R}^n$-action*. We write $\phi_t^{\boldsymbol{H}} := \phi_{t_1}^{H_1}\cdots\phi_{t_n}^{H_n}$ for this $\mathbb{R}^n$-action and $\Omega(x)$ for its orbit through $x \in X$.

Example 1.19 (Not a Hamiltonian $\mathbb{R}^n$-action) Consider the Hamiltonians x and y on $\mathbb{R}^2$. These generate an $\mathbb{R}^2$-action on $\mathbb{R}^2$ where (s,t) acts by $\phi_t^x\phi_s^y(x_0,y_0) = (x_0+s, y_0+t)$. This example is *not* a Hamiltonian $\mathbb{R}^2$-action because the Poisson bracket $\{x,y\} = 1$ is not zero (i.e. the Hamiltonians do not Poisson-commute even though the flows commute).

Remark 1.20 More generally, for a Lie group G with Lie algebra $\mathfrak{g}$, a Hamiltonian G-action is a G-action in which every one-parameter subgroup $\exp(t\xi)$, $\xi \in \mathfrak{g}$, acts as a Hamiltonian flow $\phi_t^{H_\xi}$, and the assignment $\xi \mapsto H_\xi$ is a Lie algebra map (i.e. $H_{[\xi_1, \xi_2]} = \{H_{\xi_1}, H_{\xi_2}\}$ for all $\xi_1, \xi_2 \in \mathfrak{g}$).

Definition 1.21 A submanifold L of a symplectic manifold (X, ω) is called *isotropic* if ω vanishes on vectors tangent to L and *Lagrangian* if it is isotropic and $2 \dim(L) = \dim(X)$.

Lemma 1.22 (Exercise 1.47) *If L is an isotropic submanifold of the symplectic manifold (X, ω) then $2 \dim(L) \leq \dim(X)$.*

Lemma 1.23 *Suppose that $\boldsymbol{H} : X \to \mathbb{R}^n$ generates a Hamiltonian $\mathbb{R}^n$-action. The orbits of this action are isotropic. As a consequence, if X contains a regular point[5] of $\boldsymbol{H}$ then $n \leq \frac{1}{2} \dim X$.*

Proof The tangent space to an orbit is spanned by the vectors $V_{H_1}, \ldots, V_{H_n}$, which satisfy $\omega(V_{H_i}, V_{H_j}) = \{H_i, H_j\} = 0$, so the orbits are isotropic. If $x \in X$ is a regular point then the differentials $dH_1, \ldots, dH_n$ are linearly independent at x, so the vectors $V_{H_1}(x), \ldots, V_{H_n}(x)$ span an n-dimensional isotropic space, which can have dimension at most $\frac{1}{2} \dim X$. □

Corollary 1.24 *If $\dim X = 2n$ and $\boldsymbol{H} : X \to \mathbb{R}^n$ is a smooth map with connected fibres whose components satisfy $\{H_i, H_j\} = 0$, then the regular fibres are Lagrangian orbits of the $\mathbb{R}^n$-action.*

Proof Since $\{H_i, H_j\} = 0$, Lemma 1.16 implies that H_j is constant along the flow of V_{H_i}. In particular, this means that if $x \in \boldsymbol{H}^{-1}(\boldsymbol{b})$ then its orbit $\Omega(x)$ is contained in the fibre $\boldsymbol{H}^{-1}(\boldsymbol{b})$. If $\boldsymbol{b}$ is a regular value then the fibre $\boldsymbol{H}^{-1}(\boldsymbol{b})$ is n-dimensional, and the orbit of each point in the fibre is a n-dimensional isotropic (i.e. Lagrangian) submanifold, so the fibre is a union of Lagrangian submanifolds. These orbits are open submanifolds of the fibre: if $\Omega(x) \subseteq \boldsymbol{H}^{-1}(\boldsymbol{b})$ then for any open neighbourhood $T \subseteq \mathbb{R}^n$ of 0, the subset $\{\phi_t^{\boldsymbol{H}}(x) : t \in T\}$ is an open neighbourhood of $x \in \boldsymbol{H}^{-1}(\boldsymbol{b})$ contained in $\Omega(x)$. If the fibre is connected then it cannot be a union of more than one open submanifold, so the $\mathbb{R}^n$-action is transitive on connected regular fibres, as required. □

Definition 1.25 Let (X, ω) be a $2n$-dimensional symplectic manifold. We say that a smooth map $\boldsymbol{H} : X \to \mathbb{R}^n$ is a *complete commuting Hamiltonian system* if the components $H_1, \ldots, H_n$ satisfy $\{H_i, H_j\} = 0$ for all i, j. We say that

---

[5] Recall that if $\boldsymbol{H} : X \to \mathbb{R}^n$ is a smooth map then a point $x \in X$ is called *regular* if $d\boldsymbol{H}$ is surjective at x, and a point $\boldsymbol{b} \in \mathbb{R}^n$ is called a *regular value* if the fibre $\boldsymbol{H}^{-1}(\boldsymbol{b})$ consists entirely of regular points; in this case we call $\boldsymbol{H}^{-1}(\boldsymbol{b})$ a *regular fibre*.

a complete commuting Hamiltonian system H is an *integrable Hamiltonian system* if

- $H(X)$ contains a dense open set of regular values,
- H is proper (preimages of compact sets are compact) and has connected fibres.

The first assumption rules out trivial examples; the properness condition ensures that the flows of the vector fields $V_{H_1}, \ldots, V_{H_n}$ exist for all time.

1.4 Period Lattices

We want to generalise the idea that all orbits have the same period, but now we have n Hamiltonians.

Definition 1.26 Suppose we have an integrable Hamiltonian system $H \colon X \to \mathbb{R}^n$. Let $B \subseteq H(X) \subseteq \mathbb{R}^n$ be an open subset of the image of H. A *local section over B* is a map $\sigma \colon B \to X$ such that $H \circ \sigma = \mathrm{id}$.

Remark 1.27 Note that if σ is a local section over B then $\sigma(b)$ is necessarily a regular point of H for every $b \in B$ because $dH(d\sigma(T_b B)) = \mathrm{id}(T_b B) = T_b B$.

Definition 1.28 Given an integrable Hamiltonian system $H \colon X \to \mathbb{R}^n$ and a local section $\sigma \colon B \to X$, over a subset $B \subseteq H(X)$, the *period lattice at $b \in B$* is defined to be

$$\Lambda_b^H := \{ t \in \mathbb{R}^n \ : \ \phi_t^H(\sigma(b)) = \sigma(b) \},$$

and the *period lattice* is

$$\Lambda^H := \{ (b, t) \in B \times \mathbb{R}^n \ : \ t \in \Lambda_b^H \}.$$

We will often omit the superscript H if H is clear from the context. We say that the period lattice is standard if $\Lambda = B \times (2\pi\mathbb{Z})^n$.

Lemma 1.29 Λ_b^H *consists of tuples $t \in \mathbb{R}^n$ such that ϕ_t^H fixes every point of the orbit $\Omega(\sigma(b))$.*

Proof By definition, $t \in \Lambda_b^H$ if and only if ϕ_t^H fixes $\sigma(b)$. Any other point in this orbit can be written as $\phi_{t'}^H(\sigma(b))$ for some t'. Therefore if $t \in \Lambda_b^H$, we have

$$\phi_t^H(\phi_{t'}^H(\sigma(b))) = \phi_{t'}^H(\phi_t^H(\sigma(b))) = \phi_{t'}^H(\sigma(b)),$$

so ϕ_t^H fixes every point in the orbit. $\square$

Remark 1.30   If the orbit $\Omega(\sigma(\boldsymbol{b}))$ is dense in $\boldsymbol{H}^{-1}(\boldsymbol{b})$, this means

$$\Lambda_{\boldsymbol{b}}^{\boldsymbol{H}} = \{\boldsymbol{t} \in \mathbb{R}^n \ : \ \phi_{\boldsymbol{t}}^{\boldsymbol{H}}|_{\boldsymbol{H}^{-1}(\boldsymbol{b})} = \mathrm{id}_{\boldsymbol{H}^{-1}(\boldsymbol{b})}\}.$$

Example 1.31   In Example 1.2, the Hamiltonian is $H_1(p, q) = \frac{1}{2}(p^2 + q^2)$ on $\mathbb{R}^2$. If we take $B = \mathbb{R}_{>0}$ and choose the section $\sigma(b) = (\sqrt{2b}, 0)$ then $\phi_t^{H_1}(\sigma(b)) = (\sqrt{2b}\cos t, \sqrt{2b}\sin t)$ and the period lattice is standard: every point $\sigma(b)$ returns to itself after time 2π. See Figure 1.2 (left).

Example 1.32   In Example 1.3, the Hamiltonian is $H_2(p, q) = \sqrt{p^2 + q^2}$ on $\mathbb{R}^2$. If we take the section $\sigma(b) = (b, 0)$ then we have $\phi_t^{H_2}(\sigma(b)) = (b\cos(t/b), b\sin(t/b))$ so the point $\sigma(b)$ returns to itself after time $2\pi b$. The period lattice is therefore $\{(b, 2\pi bn) \ : \ b > 0, \ n \in \mathbb{Z}\}$. See Figure 1.2 (right).

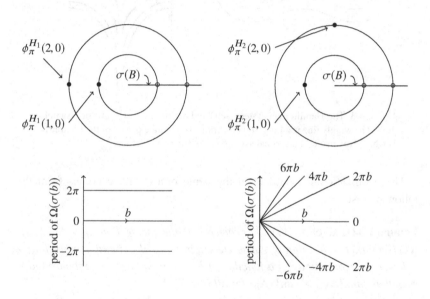

Figure 1.2 Top: Snapshots at $t = \pi$ of the Hamiltonian systems in Examples 1.31 (left) and 1.32 (right) showing the orbits and positions of $\phi_\pi^H(1, 0)$ and $\phi_\pi^H(2, 0)$. Bottom: The period lattices from Example 1.31 (left: standard) and Example 1.32 (right: non-standard).

Example 1.33   Consider a Hamiltonian system on $\mathbb{R}^2$ whose level sets are shown in Figure 1.3. This Hamiltonian generates an $\mathbb{R}$-action which has three types of orbits: the fixed points (marked $\bullet$ in the figure); the two separatrices (arcs connecting the central fixed point to itself); the remaining orbits are closed loops either inside or outside the separatrices. The separatrices have infinite period (it takes infinitely long to flow around them). If we take as a Lagrangian section the wiggly line segment on the left then the period lattice looks like the figure on the right.

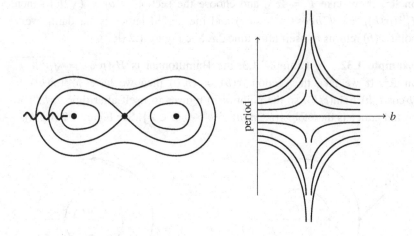

Figure 1.3 The Hamiltonian system (left) and period lattice (right) for Example 1.33. The wiggly line is a Lagrangian section. The infinite period of the separatrix is what gives rise to the vertical asymptotes of the period lattice.

The justification for 'lattice' in the name period lattice comes from the following result:

Lemma 1.34 (Exercise 1.49) *For each $b \in B$, the intersection $\Lambda_b = \Lambda \cap (\{b\} \times \mathbb{R}^n)$ is a lattice in $\mathbb{R}^n$, that is, a discrete subgroup of $\mathbb{R}^n$. The rank of the lattice is lower semicontinuous as a function of b, that is, b has a neighbourhood V such that $rank(\Lambda_{b'}) \geq rank(\Lambda_b)$ for all $b' \in V$.*

Example 1.35 In Example 1.33, the period lattice for most orbits is isomorphic to $\mathbb{Z}$, but where $\sigma(B)$ intersects the separatrix orbit the period lattice is the zero lattice; this corresponds to the vertical asymptote in Figure 1.3.

The following result can be found in Arnold's book [3, Lemma 3, p. 276], and tells us that lattices are what we think they are. We will use it below to explain why compact orbits are diffeomorphic to tori.

Lemma 1.36 *If $\Lambda \subseteq \mathbb{R}^n$ is a lattice then there is a basis $e_1, \ldots, e_n$ of $\mathbb{R}^n$ such that Λ is the $\mathbb{Z}$-linear span of the vectors $e_1, \ldots, e_k$ for some $k \leq n$.*

1.5 Liouville Coordinates

In what follows, we will usually use *Lagrangian sections* to define the period lattice, that is, sections whose image is a Lagrangian submanifold. These always exist locally:

Lemma 1.37 (Exercise 1.48)   *Let $H: X \to \mathbb{R}^n$ be an integrable Hamiltonian system. There exists a local* Lagrangian *section through any regular point x.*

Theorem 1.38 (Liouville coordinates)   *Let $H: X \to \mathbb{R}^n$ be an integrable Hamiltonian system, let $B \subseteq \mathbb{R}^n$ be an open set, and let $\sigma: B \to X$ be a local Lagrangian section. Define*

$$\Psi: B \times \mathbb{R}^n \to X, \qquad \Psi(\boldsymbol{b}, \boldsymbol{t}) = \phi_{\boldsymbol{t}}^{\boldsymbol{H}}(\sigma(\boldsymbol{b})).$$

Then Ψ is both an immersion and a submersion and $\Psi^* \omega = \sum db_i \wedge dt_i$, where $(b_1, \ldots, b_n)$ are the standard coordinates on $B \subseteq \mathbb{R}^n$. This means that $(b_1, \ldots, b_n, t_1, \ldots, t_n)$ provide local symplectic coordinates on a neighbourhood of $\sigma(B)$; we call these Liouville coordinates.

Proof   We first verify that $\Psi^* \omega = \sum_{i=1}^{n} db_i \wedge dt_i$ on pairs of basis vectors ∂_{b_i} and ∂_{t_i}. First, observe that, by definition of Ψ, we have

$$\Psi_* \partial_{b_i} = (\phi_{\boldsymbol{t}}^{\boldsymbol{H}})_* \sigma_* (\partial_{b_i}), \qquad \Psi_* \partial_{t_i} V_{H_i}.$$

The vectors $\Psi_* \partial_{b_i}$ and $\Psi_* \partial_{b_j}$ are tangent to $\phi_{\boldsymbol{t}}^{\boldsymbol{H}}(\sigma(B))$, which is the image of a Lagrangian submanifold under a series of Hamiltonian flows, hence Lagrangian. Therefore $\omega(\Psi_* \partial_{b_i}, \Psi_* \partial_{b_j}) = 0$.

Since $\Psi_* \partial_{t_i} = V_{H_i}$, we have $\omega(\Psi_* \partial_{t_i}, \Psi_* \partial_{t_j}) = \omega(V_{H_i}, V_{H_j}) = \{H_i, H_j\} = 0$.

Finally, we have $\omega(\Psi_* \partial_{b_i}, \Psi_* \partial_{t_j}) = -(\iota_{V_{H_j}} \omega)(\Psi_* \partial_{b_i}) dH_j(\Psi_* \partial_{b_i})$. Since the flow along $\phi_{\boldsymbol{t}}^{\boldsymbol{H}}$ preserves the level sets of H_j, we have $(H_j \circ \Psi)(\boldsymbol{b}, \boldsymbol{t}) = b_j$. Therefore $dH_j(\Psi_*(\partial_{b_i})) = db_j(\partial_{b_i}) = \delta_{ij}$. This completes the verification that $\Psi^* \omega = \sum db_i \wedge dt_i$.

This implies that Ψ is both an immersion and a submersion: if this failed at some point then $\Psi^* \omega$ would be degenerate there.   □

Remark 1.39   Note that the period lattice is given by $\Lambda^{\boldsymbol{H}} = \Psi^{-1}(\sigma(B))$. Since Ψ is a symplectic map and σ is a Lagrangian section, the period lattice is a Lagrangian submanifold of $B \times \mathbb{R}^n$ with respect to $\sum db_i \wedge dt_i$.

1.6 The Arnold–Liouville Theorem

Theorem 1.40 (Little Arnold–Liouville theorem)   *Let $H: X \to \mathbb{R}^n$ be an integrable Hamiltonian system and $\sigma: B \to X$ be a local section. Each orbit*

$\Omega(\sigma(\boldsymbol{b}))$ *is diffeomorphic to* $\left(\mathbb{R}^k/\mathbb{Z}^k\right) \times \mathbb{R}^{n-k}$ *for some* k. *In particular, if* $\Omega(\sigma(\boldsymbol{b}))$ *is compact then it is a torus.*

Proof The action of $\mathbb{R}^n$ defines a diffeomorphism $\mathbb{R}^n/\Lambda_b \to \Omega(\sigma(\boldsymbol{b}))$. Since Λ_b is a lattice, the result follows from the classification of lattices in Lemma 1.36. □

We now focus attention on a neighbourhood of a regular fibre (i.e. one containing no critical points). By Corollary 1.24, a regular fibre is an orbit of the $\mathbb{R}^n$-action. Since $\boldsymbol{H}$ is proper, its fibres are compact, so by Theorem 1.40, a regular fibre is a torus; this is the analogue of assuming that our fibres are circles in Theorem 1.4. Since the set of regular values is open, we can shrink the domain B of our local Lagrangian section so that it is a disc consisting entirely of regular values. Our goal is to find a map $\alpha \colon B \to \mathbb{R}^n$ such that $\alpha \circ \boldsymbol{H}$ has standard period lattice.

Lemma 1.41 (Exercise 1.50) *Let* $\boldsymbol{H} \colon X \to B \subseteq \mathbb{R}^n$ *be an integrable Hamiltonian system over a disc with only regular fibres, let* $\alpha \colon B \to C \subseteq \mathbb{R}^n$ *be a diffeomorphism, and let* $\boldsymbol{G} := \alpha \circ \boldsymbol{H}$. *Let* $A(b)$ *be the matrix with* ijth *entry*[6] $A_{ij}(b) = \frac{\partial \alpha_i}{\partial b_j}(b)$ *(the Jacobian of* α). *Then:*

(i) the Hamiltonian vector fields of $\boldsymbol{G}$ *and* $\boldsymbol{H}$ *are related by* $V_{G_i} = \sum_j A_{ij} V_{H_j}$,

(ii) the Hamiltonian flows of $\boldsymbol{G}$ *and* $\boldsymbol{H}$ *are related by* $\phi_t^G = \phi_{A^T t}^H$, *and*

(iii) the period lattices Λ^G *and* Λ^H *are related by* $A^T \Lambda_{\alpha(b)}^G = \Lambda_b^H$.

Theorem 1.42 (Action-angle coordinates) *Let* $\boldsymbol{H} \colon X \to B \subseteq \mathbb{R}^n$ *be an integrable Hamiltonian system over the disc with only regular fibres and pick a local Lagrangian section* σ. *There is a local change of coordinates* $\alpha \colon B \to C \subseteq \mathbb{R}^n$ *such that* $\boldsymbol{G} := \alpha \circ \boldsymbol{H}$ *generates a Hamiltonian torus action on* X. *In other words, the period lattice* Λ^G *is standard and the map* $(\boldsymbol{c}, \boldsymbol{t}) \mapsto \phi_t^G(\sigma(\alpha^{-1}(\boldsymbol{c})))$ *defined in Theorem 1.38 descends to give a symplectomorphism* $C \times (\mathbb{R}/2\pi\mathbb{Z})^n \to \boldsymbol{G}^{-1}(C) = \boldsymbol{H}^{-1}(B)$.

Proof The following proof is due to Duistermaat [26].

For each $\boldsymbol{b} \in B$, let $2\pi\tau_1(\boldsymbol{b}), \ldots, 2\pi\tau_n(\boldsymbol{b}) \in \mathbb{R}^n$ be a collection of vectors (smoothly varying in $\boldsymbol{b}$) which span the lattice of periods Λ_b^H. This is possible because B is contractible so there is no obstruction to picking sections of the projection $\Lambda \to B$. This means that $\phi_{\tau_i(b)}^H = \mathrm{id}$ for $i = 1, \ldots, n$. Let us write $\tau_i(\boldsymbol{b}) = (A_{i1}(\boldsymbol{b}), \ldots, A_{in}(\boldsymbol{b}))$. Let A be the matrix with ijth entry $A_{ij}(\boldsymbol{b})$. Then $\Lambda_b^H = 2\pi A^T \mathbb{Z}^n$. By Lemma 1.41(iii), it is sufficient to find a map $\alpha = (\alpha_1, \ldots, \alpha_n) \colon B \to \mathbb{R}^n$ whose Jacobian $\partial \alpha_i/\partial b_j$ is A_{ij}.

---

[6] I.e. ith row, jth column.

By the Poincaré lemma, we can find such functions α_i provided

$$\frac{\partial A_{ij}}{\partial b_k} = \frac{\partial A_{ik}}{\partial b_j}, \tag{1.2}$$

so it remains to check this identity.

Let $\Psi: B \times \mathbb{R}^n \to X$ be the Liouville coordinates associated to our choice of Lagrangian section and $\Lambda = \Psi^{-1}(\sigma(B))$ be the period lattice. Since Ψ is symplectic and $\sigma(B)$ is Lagrangian, Λ is Lagrangian. Moreover, Λ is a union of sheets, each traced out by a single lattice point. For example, $\{(b, \tau_i(b)) : b \in B\}$ traces out a Lagrangian sheet for each i. In coordinates, this is $\{(b_1, \ldots, b_n, A_{i1}(b), \ldots, A_{in}(b)) : b \in B\}$, which is Lagrangian if and only if Equation (1.2) holds (Exercise 1.51). □

Definition 1.43 The Liouville coordinates associated to the new, periodic Hamiltonian system are called *action-angle coordinates*. More precisely, the new Hamiltonians $\alpha_1 \circ H, \ldots, \alpha_n \circ H$ are called *action coordinates* and the new 2π-periodic conjugate coordinates $t_1, \ldots, t_n$ are called *angle coordinates*.

Corollary 1.44 (Big Arnold–Liouville theorem) *If $H: M \to \mathbb{R}^n$ is an integrable Hamiltonian system then any regular fibre is a torus and admits a neighbourhood symplectomorphic to $B \times T^n$, where $B \subseteq \mathbb{R}^n$ is an open ball and the symplectic form is given by $\sum_{i=1}^n db_i \wedge dt_i$. Under this symplectomorphism, the orbits of the original system are sent to the tori $\{b\} \times T^n$.*

1.7 Solutions to Inline Exercises

Exercise 1.45 (Remark 1.15) *Recall that the flows along two vector fields commute if and only if the Lie bracket of the vector fields vanishes. Show that two Hamiltonian flows ϕ_t^F and ϕ_t^G commute if and only if the Poisson bracket $\{F, G\}$ is locally constant.*

Solution We have $[V_F, V_G] = V_{\{F,G\}}$ by Lemma 1.13. Since $\iota_{V_{\{F,G\}}} \omega = -d\{F, G\}$, we see that the Lie bracket vanishes if and only if $d\{F, G\} = 0$ so that all partial derivatives of $\{F, G\}$ vanish. This happens if and only if $\{F, G\}$ is locally constant. □

Exercise 1.46 (Lemma 1.16) *Let F and G be smooth functions. Define $F_t(x) := F(\phi_t^G(x))$. Then $\frac{dF_t}{dt} = \{G, F_t\}$.*

Proof We have $\frac{dF_t}{dt} = dF(V_G) = -\omega(V_F, V_G) = \{G, F\}$. □

Exercise 1.47 (Lemma 1.22)   *If L is an isotropic submanifold of the symplectic manifold (X, ω) then $2 \dim(L) \leq \dim(X)$.*

Proof   For any point $x \in L$, the tangent space $T_x L$ is an isotropic subspace of the symplectic vector space $T_x X$. The claim now follows from Lemma A.7 in the appendix on symplectic linear algebra.   □

Exercise 1.48 (Lemma 1.37)   *Let $H: X \to \mathbb{R}^n$ be an integrable Hamiltonian system. There exists a local Lagrangian section through any regular point x.*

Proof   It is a theorem of Darboux (see [3, Section 43.B], [5, Corollary I.1.11], [77, Theorem 3.15]) that any point x in a symplectic manifold is the centre of a coordinate chart $(p_1, \ldots, p_n, q_1, \ldots, q_n)$ where the symplectic form is $\sum_i dp_i \wedge dq_i$. Let us work locally in these coordinates. We treat this local chart as a symplectic vector space and use some notions from the appendix on symplectic linear algebra. If we define $J: \mathbb{R}^{2n} \to \mathbb{R}^{2n}$ to be the linear map $J(p, q) = (-q, p)$ then J is an ω-compatible complex structure (see Definition A.9). Thus, if L is a Lagrangian subspace in $\mathbb{R}^{2n}$, the subspace JL is a complementary Lagrangian subspace (Lemma A.13). Since $T_x \Omega(x)$ is the tangent space to the orbit $\Omega(x)$, its image $JT_x \Omega(x)$ is a Lagrangian complement. The subspace $JT_x \Omega(x)$ is the tangent space of a linear Lagrangian submanifold L of the Darboux ball, which is transverse to $\Omega(x)$ at x. The differentials $dH_1, \ldots, dH_n$ are linearly independent at x but vanish on $T_x \Omega(x)$ because H is constant on $\Omega(x)$. Therefore these differentials restrict to linearly independent forms on L near x. This implies that the map $H|_L: L \to \mathbb{R}^n$ is a local diffeomorphism in a neighbourhood of x, so that its local inverse is a local section of H near x whose image is contained in L and hence Lagrangian.   □

Exercise 1.49 (Lemma 1.34)   *Let $H: X \to \mathbb{R}^n$ be an integrable Hamiltonian system, let $B \subseteq \mathbb{R}^n$ be an open set of regular values and let $\sigma: B \to X$ be a local Lagrangian section; write Λ for the period lattice. For each $b \in B$, the intersection $\Lambda_b = \Lambda \cap (\{b\} \times \mathbb{R}^n)$ is a lattice in $\mathbb{R}^n$, that is, a discrete subgroup of $\mathbb{R}^n$. The rank of the lattice is lower semicontinuous as a function of b, that is, b has a neighbourhood V such that $rank(\Lambda_{b'}) \geq rank(\Lambda_b)$ for all $b' \in V$.*

Proof   Let $\sigma: B \to X$ be a local Lagrangian section of H such that $\sigma(b)$ is a regular point of H for all $b \in B$. We will first show that, for all $b \in B$, the period lattice Λ_b is a discrete subgroup of $\mathbb{R}^n$.

The subset Λ_b is the stabiliser of b under the action of $\mathbb{R}^n$, so it is a subgroup of $\mathbb{R}^n$. To prove discreteness, we need to show that there is an open set $W \subseteq \mathbb{R}^n$ such that $W \cap \Lambda_b = \{0\}$. Since $\Psi: B \times \mathbb{R}^n \to X$ is a local diffeomorphism, there is an open set $W' \subseteq B \times \mathbb{R}^n$ (containing $(b, 0)$) such that $\Psi: W' \to \Psi(W')$

is a diffeomorphism. There exist open sets $b \in W_1 \subseteq B$ and $0 \in W_2 \subseteq \mathbb{R}^n$ such that $W_1 \times W_2 \subseteq W'$, as these product sets form a basis for the product topology. In particular, 0 is the only point t in W_2 such that $\Psi(b, t) = b$. We may therefore take $W = W_2$ to see that Λ_b is discrete.

To see that the rank of the lattice is lower semicontinuous, we need to show, for each $b \in B$, there is a neighbourhood V of b such that $rank(\Lambda_{b'}) \geq rank(\Lambda_b)$ for $b' \in V$.

Let $\lambda_1(b), \ldots, \lambda_k(b)$ be a $\mathbb{Z}$-basis for $\Lambda_b = \{t \in \mathbb{R}^n : \phi_t^H(b) = b\}$. Then, since Ψ is an immersion (Theorem 1.38), there is an open neighbourhood of $b \in B$ such that, for b' in this open neighbourhood, there are solutions $t = \lambda_1(b'), \ldots, t = \lambda_k(b')$ to the equation $\phi_t^H(b') = b'$ which vary continuously in b'. Since the condition of being linearly independent is an open condition, the points $\lambda_1(b'), \ldots, \lambda_k(b')$ are linearly independent for b' in a, possibly smaller, neighbourhood of b, so the rank of the lattice $\Lambda_{b'}$ is at least k for b' in a neighbourhood of b. $\quad\square$

Exercise 1.50 (Lemma 1.41) *Let $H \colon X \to B \subseteq \mathbb{R}^n$ be an integrable Hamiltonian system over a disc with only regular fibres, let $\alpha \colon B \to C \subseteq \mathbb{R}^n$ be a diffeomorphism, and let $G := \alpha \circ H$. Let $A(b)$ be the matrix with ijth entry[7] $A_{ij}(b) = \frac{\partial \alpha_i}{\partial b_j}(b)$ (the Jacobian of α). Then:*

  (i) the Hamiltonian vector fields of G and H are related by $V_{G_i} = \sum_j A_{ij} V_{H_j}$,
  (ii) the Hamiltonian flows of G and H are related by $\phi_t^G = \phi_{A^T t}^H$, and
  (iii) the period lattices Λ^G and Λ^H are related by $A^T \Lambda_{\alpha(b)}^G = \Lambda_b^H$.

Solution Let us write $A_{ij} = \frac{\partial \alpha_i}{\partial b_j}$. We have

$$\iota_{\sum_j A_{ij} V_{H_j}} \omega = \sum_j \frac{\partial \alpha_i}{\partial b_j} \iota_{V_{H_j}} \omega$$

$$= -\sum_j \frac{\partial \alpha_i}{\partial b_j} dH_j$$

$$= -d(\alpha_i \circ H) = -dG_i.$$

This proves (i): $V_{G_i} = \sum_j A_{ij} V_{H_j}$. Thus, if t is the row vector $(t_1, \ldots, t_n)$, then

$$\sum_i t_i V_{G_i} = \sum_{i,j} t_i A_{ij} V_{H_j},$$

where the matrix A_{ij} is constant on each orbit. Therefore, we obtain (ii): $\phi_t^G = \phi_{A^T t}^H$.

---

[7] I.e. ith row, jth column.

The lattice $\Lambda^G_{\alpha(b)}$ of G consists of tuples $t = (t_1, \ldots, t_n)$ such that $\phi^G_t = \mathrm{id}$ on $G^{-1}(\alpha(b))$. By (ii), this is equivalent to $\phi^H_{A^T t} = \mathrm{id}$ on the orbit $H^{-1}(b)$, so $t \in \Lambda^G_{\alpha(b)}$ if and only if $A^T t \in \Lambda^H_b$, which gives (iii):

$$A^T \Lambda^G_{\alpha(b)} = \Lambda^H_b. \qquad \qquad \square$$

Exercise 1.51 (From proof of Theorem 1.42) *Show that a section $\sigma(b) = (b, t(b))$ is Lagrangian with respect to the symplectic form $\omega = \sum db_i \wedge dt_i$ if and only if $\partial t_i / \partial b_j = \partial t_j / \partial b_i$ for all i, j.*

Solution The tangent space to the section σ is spanned by the vectors $\sigma_*(\partial_{b_i})$ so it suffices to check that $\omega(\sigma_*(\partial_{b_i}), \sigma_*(\partial_{b_j})) = 0$ for all i, j. We have $\sigma_*(\partial_{b_i}) = \partial_{b_i} + \sum_k (\partial t_i / \partial b_k) \partial_{b_k}$, which gives

$$\omega(\sigma_*(\partial_{b_i}), \sigma_*(\partial_{b_j})) = \partial t_i / \partial b_j - \partial t_j / \partial b_i. \qquad \qquad \square$$

2

Lagrangian Fibrations

We have seen that an integrable Hamiltonian system is a map $X \to \mathbb{R}^n$ whose regular fibres are Lagrangian submanifolds. This structure, called a *Lagrangian fibration*,[1] turns out to be very useful for studying the geometry and topology of symplectic manifolds.

In this chapter, we introduce a general definition of Lagrangian fibration. We then discuss the *regular Lagrangian fibrations*: those with no critical points, namely proper submersions $X \to B$ with connected Lagrangian fibres. We will see that these are locally the same as integrable Hamiltonian systems (Remark 2.7). In particular, the fibres are tori (Corollary 2.8). For this reason, we often use the name *Lagrangian torus fibration* instead of Lagrangian fibration. Next, we will see that local action coordinates equip the image B with a geometric structure called an *integral affine structure*, which can also be understood in terms of the symplectic areas of cylinders connecting fibres. Finally, we will show that under certain assumptions (existence of a global Lagrangian section), the integral affine manifold B is enough information to reconstruct the Lagrangian fibration $X \to B$ completely.

As the book progresses, we will allow our fibrations to have progressively worse critical points.

2.1 Lagrangian Fibrations

Definition 2.1 Recall that a stratification of a topological space B is a filtration

$$\emptyset =: B_{-1} \subseteq B_0 \subseteq \cdots \subseteq B_d \subseteq B_{d+1} \subseteq \cdots \subseteq B,$$

---

[1] The word *fibration* also appears in algebraic topology (e.g. *Serre fibrations*), where it describes maps with a homotopy lifting property. Lagrangian torus fibrations are not fibrations in that sense: though they are fibre bundles over the regular locus, homotopy lifting fails near the critical points. This is an unfortunate accident of history.

where each B_d is a closed subset such that, for each d, the d-stratum $S_d(B) :=$ $B_d \setminus B_{d-1}$ is a smooth d-dimensional manifold (possibly empty) and $B = \bigcup_{d \geq 0} B_d$. We say that B is finite-dimensional if the d-stratum is empty for sufficiently large d, and we say that B is n-dimensional if B is finite-dimensional and n is maximal such that $S_n(B)$ is nonempty (in this case we call $S_n(B)$ the *top stratum*).

We adopt the following working definition of a Lagrangian torus fibration, given in [35, Definition 2.5]. It is extremely weak because it places no restrictions on the critical points of the fibration.

Definition 2.2 Let (X, ω) be a $2n$-dimensional symplectic manifold and B be an n-dimensional stratified space. A Lagrangian torus fibration $f \colon X \to B$ is a proper continuous map such that f is a smooth submersion over the top stratum with connected Lagrangian fibres, and the other fibres are themselves connected stratified spaces with isotropic strata. We call $B^{reg} := S_n(B)$ the *regular locus* of H and $B^{sing} := B \setminus S_n(B)$ the *discriminant locus*.

Remark 2.3 Throughout Chapter 1, B denoted an open subset of $\mathbb{R}^n$. This is no longer the case. However, it is still the target ('base') of the fibration, hence the choice of letter.

2.2 Regular Lagrangian Fibrations

We first study Lagrangian fibrations with no critical points. It turns out (Lemma 2.6) that these are locally equivalent to integrable Hamiltonian systems.

Definition 2.4 We say that a Lagrangian fibration $f \colon X \to B$ is *regular* if $B = B^{reg}$, that is, if f is a smooth proper submersion with connected Lagrangian fibres.

Lemma 2.5 *Let (X, ω) be a symplectic manifold. Suppose that $H \colon X \to \mathbb{R}$ is a Hamiltonian function and $L \subseteq X$ is a Lagrangian submanifold such that $L \subseteq H^{-1}(c)$ for some $c \in \mathbb{R}$. Then $\phi_t^H(x) \in L$ for all $x \in L$, $t \in \mathbb{R}$. That is, L is invariant under the Hamiltonian flow of H.*

Proof Since $L \subseteq H^{-1}(c)$, the function H is constant on L, so the directional derivative $v(H) = dH(v)$ vanishes whenever $v \in TL$. We have $\iota_{V_H} \omega = -dH$. If $v \in TL$ then

$$\omega(V_H, v) = -dH(v) = 0.$$

This means that V_H is in the symplectic orthogonal complement[2] $(TL)^\omega$. Since L is Lagrangian, $TL = (TL)^\omega$, so this shows that $V_H \in TL$. Since V_H is tangent to L, the flow of V_H preserves L.   □

Lemma 2.6 Let (X, ω) be a symplectic $2n$-manifold, B be an n-manifold, and $f : X \to B$ be a regular Lagrangian fibration. Let $(b_1, \ldots, b_n)$ be local coordinates on B. The functions $b_1 \circ f, \ldots, b_n \circ f$ Poisson commute.

Proof Fix a point $c \in B$ with $b_i(c) = c_i$. The Lagrangian fibre $f^{-1}(c)$ is contained in all the level sets $\{b_i \circ f = c_i\}$, $i = 1, \ldots, n$. By Lemma 2.5, the Hamiltonian vector field $V_{b_i \circ f}$ is tangent to L (for all i). Therefore,

$$\{b_i \circ f, b_j \circ f\} = \omega(V_{b_i \circ f}, V_{b_j \circ f}) = 0,$$

because $V_{b_i \circ f}, V_{b_j \circ f} \in TL$ and L is Lagrangian.   □

Remark 2.7 In particular, f is locally modelled on an integrable Hamiltonian system.

Corollary 2.8 (Exercise 2.35) *If $f : X \to B$ is a proper submersion with connected Lagrangian fibres then the fibres are Lagrangian tori.*

2.3 Integral Affine Structures

The big Arnold–Liouville theorem (Corollary 1.44) gives us more information than Corollary 2.8: we will be able to show that the base of the Lagrangian fibration has an *integral affine structure*.

Definition 2.9 An integral affine transformation is a map $T : \mathbb{R}^n \to \mathbb{R}^n$ of the form[3] $T(b) = bA + C$ where $A \in GL(n, \mathbb{Z})$ and $C \in \mathbb{R}^n$. An integral affine structure on a manifold B is an atlas for B whose transition functions are integral affine transformations.

Lemma 2.10 *Suppose $G : X \to \mathbb{R}^n$ and $H : X \to \mathbb{R}^n$ are submersions defining integrable Hamiltonian systems such that the period lattices are both standard. Suppose that $\psi : H(X) \to G(X)$ is a diffeomorphism such that $G = \psi \circ H$. Then ψ is (the restriction to $H(X)$ of) an integral affine transformation.*

Proof Let $\phi_t^G = \phi_{t_1}^{G_1} \cdots \phi_{t_n}^{G_n}$ and $\phi_t^H = \phi_{t_1}^{H_1} \cdots \phi_{t_n}^{H_n}$ be the Hamiltonian $\mathbb{R}^n$-actions. Since $G = \psi \circ H$, Lemma 1.41(iii) implies $A(b)\Lambda_{\psi(b)}^G = \Lambda_b^H$, where $A(b) = d_b\psi$. Since both period lattices are assumed to be standard, this means

---

[2] See Definition A.3 for the definition of the symplectic orthogonal complement.
[3] We think of $\mathbb{R}^n$ as consisting of row vectors and matrices acting on the right.

$A(b) \in GL(n, \mathbb{Z})$ for all $b \in G(X)$. Since $GL(n, \mathbb{Z})$ is discrete, this is only possible if $d\psi$ is constant. Thus $\psi(b) = bA + C$ for some $A \in GL(n, \mathbb{Z})$ and $C \in \mathbb{R}^n$.   □

Remark 2.11   This proof contains the first instance of a useful trick we will use repeatedly in what follows. Namely, by showing that the derivative of ψ belongs to some discrete set, we were able to severely constrain ψ. For further examples of this trick in action, see Proposition 3.3 (the boundary of the moment polytope is piecewise linear) and Theorem 5.1 ('visible Lagrangians' live over straight lines).

Theorem 2.12   *If $f: X \to B$ is a regular Lagrangian fibration then B inherits an integral affine structure.*

Proof   Suppose we are given a coordinate chart[4] $\varphi: B \dashrightarrow \mathbb{R}^n$. By Lemma 2.6, $\varphi \circ f$ is an integrable Hamiltonian system. Let $\alpha: \mathbb{R}^n \dashrightarrow \mathbb{R}^n$ be the map constructed in the proof of Theorem 1.42 so that $\alpha \circ \varphi \circ f$ are action coordinates. This gives us a modified chart $\alpha \circ \varphi: B \dashrightarrow \mathbb{R}^n$. If we modify a whole atlas in this way, we obtain a new atlas; we will check that the resulting transition functions are integral affine transformations. Suppose we have charts $\varphi_1: B \dashrightarrow \mathbb{R}^n$ and $\varphi_2: B \dashrightarrow \mathbb{R}^n$ which we modify using $\alpha_1: \mathbb{R}^n \dashrightarrow \mathbb{R}^n$, $\alpha_2: \mathbb{R}^n \dashrightarrow \mathbb{R}^n$. The transition map for the modified atlas is $\psi_{12} := \alpha_2 \circ \varphi_2 \circ \varphi_1^{-1} \circ \alpha_1^{-1}$. We know that $H := \alpha_1 \circ \varphi_1 \circ f$ and $G := \alpha_2 \circ \varphi_2 \circ f$ are integrable systems with standard period lattice, and $\psi_{12} \circ H = G$, so by Lemma 2.10, ψ_{12} is an integral affine transformation.   □

Remark 2.13   In the construction of this integral affine structure, we modified the atlas and, hence, the smooth structure of B. In other words, we don't get to pick the smooth structure on B: it is dictated to us by the geometry of the fibration.

2.4 Flux Map

There is a more geometric way to characterise the action coordinates. Let $f: X \to B$ be a regular Lagrangian fibration. We assume, for simplicity,[5] that $\omega = d\lambda$ for some 1-form λ.

Consider the local system $\xi \to B$ whose fibre over b is the abelian group

---

[4]   We write partially-defined maps with $\dashrightarrow$ to save overburdening the notation with domains and targets.

[5]   (Exercise 2.36) Explain how to modify the construction to get an integral affine structure on B, even if ω is not exact. Disclaimer: this is one of the exercises that requires a lot of work.

$H_1(f^{-1}(b); \mathbb{Z}) \cong \mathbb{Z}^n$. Let $p \colon \tilde{B} \to B$ be the universal cover and let $\tilde{\xi} = p^*\xi$. Since $\tilde{B}$ is simply-connected, $\tilde{\xi}$ is trivial. Let $c_1, \ldots, c_n$ be a $\mathbb{Z}$-basis of continuous sections of $\tilde{\xi} \to \tilde{B}$.

Definition 2.14 (Flux map) The *flux map* is defined to be the map $I \colon \tilde{B} \to \mathbb{R}^n$ given by

$$I(\tilde{b}) = (I_1(\tilde{b}), \ldots, I_n(\tilde{b})) := \left(\frac{1}{2\pi} \int_{c_1(\tilde{b})} \lambda, \ldots, \frac{1}{2\pi} \int_{c_n(\tilde{b})} \lambda \right).$$

Lemma 2.15 (Flux map = action coordinates) *Suppose that $\tilde{U} \subseteq \tilde{B}$ and $U \subseteq B$ are open subsets such that $p|_{\tilde{U}} \colon \tilde{U} \to U$ is a diffeomorphism. Then $I \circ (p|_{\tilde{U}})^{-1} \colon U \to \mathbb{R}^n$ gives action coordinates on U.*

Proof By Corollary 1.44, it is sufficient to prove this for the local model $(U \times T^n, \omega_0)$ where $\omega_0 = \sum db_i \wedge dt_i)$. In that case, we can pick $\lambda = \sum b_i dt_i$ and take $c_1, \ldots, c_n$ to be the standard basis of $H_1(T^n; \mathbb{Z})$. Then we get $I_i(b) = b_i$, which recovers the action coordinates. $\square$

Definition 2.16 (Fundamental action domain) We call $I(\tilde{U})$ a *fundamental action domain* for the Lagrangian fibration.

Remark 2.17 If we pick a different λ' such that $d\lambda' = d\lambda$ then $\lambda - \lambda'$ is closed, so $\int_{c_i(b)} (\lambda - \lambda')$ is constant (by Stokes's theorem) and the flux map changes by an additive constant. If we pick a different $\mathbb{Z}$-basis $(c'_1, \ldots, c'_n)$ then we can express the new integrals as a $\mathbb{Z}$-linear combination of $I_1, \ldots, I_n$. This means that the flux map is determined up to an integral affine transformation.

The integral affine structure from Theorem 2.12 can now be understood in the following way. We pull back the integral affine structure from $\mathbb{R}^n$ along I to get an integral affine structure on $\tilde{B}$; this integral affine structure on $\tilde{B}$ descends to one on B (it is invariant under the action of deck transformations).[6] We will prove this because it introduces an important new idea: the *affine monodromy*.

Corollary 2.18 *If we equip $\tilde{B}$ with the integral affine structure pulled back from $\mathbb{R}^n$ along I then it is invariant under the action of the deck group of the cover $p \colon \tilde{B} \to B$.*

Proof If $g \colon \tilde{B} \to \tilde{B}$ is a deck transformation of the cover p then

$$c_1(\tilde{b}), \ldots, c_n(\tilde{b}) \quad \text{and} \quad c_1(\tilde{b}g), \ldots, c_n(\tilde{b}g)$$

are both $\mathbb{Z}$-bases for the $\mathbb{Z}$-module $H_1(f^{-1}(p(\tilde{b})); \mathbb{Z})$ and therefore they are

---

[6] Conventions: we think of $I(b)$ as a row vector, write concatenation of loops as $\alpha \cdot \beta$ meaning 'follow α then β', and write the deck group acting on the right.

related by some change-of-basis matrix $M(g) \in GL(n, \mathbb{Z})$. This implies that $I(\tilde{b}g) = I(\tilde{b})M(g)$. Since $M(g)$ is an integral affine transformation, this shows that the integral affine structure descends to the quotient B. $\square$

Note that, with our conventions, $M(g_1 g_2) = M(g_1)M(g_2)$. Indeed, the homomorphism $M \colon \pi_1(B) \to GL(n, \mathbb{Z})$ is the monodromy of the local system $\xi \to B$.

Definition 2.19 We call $M \colon \pi_1(B) \to GL(n, \mathbb{Z})$ the *affine monodromy* in what follows. The first example we will encounter where the affine monodromy is nontrivial will be the fibrations with focus–focus critical points in Chapter 6.

Remark 2.20 The manifold B can be reconstructed in the usual way as a quotient of a closed fundamental domain for the universal cover $\tilde{B} \to B$ where the identifications are made using deck transformations. If we wish to reconstruct the integral affine structure on B then we use a fundamental action domain and the identifications are made using the integral affine transformations $M(g)$ corresponding to deck transformations g.

Remark 2.21 Given any integral affine manifold B, there is a *developing map*, that is, a (globally defined) local diffeomorphism $I \colon \tilde{B} \to \mathbb{R}^n$ from the universal cover into Euclidean space such that the integral affine structure inherited by $\tilde{B}$ from the covering map agrees with the pullback of the integral affine structure along the developing map. In our context, the flux map is the developing map.

Remark 2.22 Suppose that $f \colon X \to B$ is an integrable system with $B \subseteq \mathbb{R}^n$, so that B already has an integral affine structure as an open subset of $\mathbb{R}^n$. This does not agree with the integral affine structure constructed in Corollary 2.18 unless the period lattice is standard.

2.5 Uniqueness

Definition 2.23 Let $f \colon X \to B$ and $g \colon Y \to C$ be regular Lagrangian fibrations. If $\phi \colon B \to C$ is a diffeomorphism then a *symplectomorphism fibred over* ϕ is a symplectomorphism $\Phi \colon X \to Y$ such that $g \circ \Phi = \phi \circ f$.

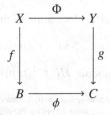

If $\phi = \mathrm{id}$, we will simply call Φ a *fibred symplectomorphism*, and if, moreover, $f = g$ then we call Φ a *fibred automorphism of f*.

An argument similar to the one which proved Lemma 2.10 shows that the map ϕ is an isomorphism of integral affine manifolds $B \to C$. We now tackle the converse question: if there is an integral affine isomorphism $\phi \colon B \to C$, is there a symplectomorphism $X \to Y$ fibred over ϕ? We first prove some preliminary lemmas.

Lemma 2.24 *Let $\Phi \colon X \to X$ be a fibred automorphism of $f \colon X \to B$ and suppose there is a Lagrangian section $\sigma \colon B \to X$ such that $\Phi \circ \sigma = \sigma$. Then $\Phi = \mathrm{id}$.*

Proof The property that $\Phi = \mathrm{id}$ can be checked locally, so we lose nothing by passing to a small affine coordinate chart in B. Without loss of generality, therefore, we will assume that $f = H \colon X \to B \subseteq \mathbb{R}^n$ is an integrable Hamiltonian system with Lagrangian section σ. By Corollary 1.44, $X \cong B \times T^n$ with symplectic form $\omega = \sum db_i \wedge dt_i$. Since we have used the section σ to define the Liouville coordinates, the section is given in these coordinates by $\sigma(b) = (b, 0)$. The fact that Φ is fibred means that $\Phi(b, t) = (b, q(b, t))$ for some function $q(b, t)$. The condition that Φ is symplectic means in particular that $\omega(\Phi_* \partial_{b_i}, \Phi_* \partial_{t_j}) = \delta_{ij}$, which becomes $\partial q_i / \partial t_j = \delta_{ij}$. Upon integrating, this means $q(b, t) = q(b, 0) + t$, so the condition $q(b, 0) = 0$ tells us that $q(b, t) = t$, and hence Φ is the identity. $\qquad\square$

Lemma 2.25 *Assume that $F \colon X \to \mathbb{R}^n$ and $G \colon Y \to \mathbb{R}^n$ are integrable Hamiltonian systems with no critical points. Assume that the period lattices Λ^F and Λ^G are both standard, and that we are given global Lagrangian sections σ of F and τ of G. Suppose there is an integral affine transformation $\phi \colon \mathbb{R}^n \to \mathbb{R}^n$ such that $\phi(F(X)) = G(X)$. Then there is a unique symplectomorphism $\Phi \colon X \to Y$ fibred over ϕ satisfying $\Phi \circ \sigma = \tau \circ \phi$.*

Proof Write $F = (F_1, \ldots, F_n)$ and $G = (G_1, \ldots, G_n)$. Let $(s_1, \ldots, s_n)$ and $(t_1, \ldots, t_n)$ be the 2π-periodic Liouville (angle) coordinates associated to the Lagrangian sections. Write $\phi(b) = bA + C$ for some $A \in GL(n, \mathbb{Z})$ and $C \in \mathbb{R}^n$. As usual, we think of F and G as row vectors and write A acting on the right.

By Corollary 1.44, X is symplectomorphic to $F(X) \times T^n$ with symplectic form $\sum_i dF_i \wedge ds_i$ and Y is symplectomorphic to $G(X) \times T^n$ with symplectic form $\sum_i dG_i \wedge dt_i$. Under these identifications, we have $\sigma(b) = (b, 0)$ and $\tau(c) = (c, 0)$.

Define a map $F(X) \times \mathbb{R}^n \to G(X) \times \mathbb{R}^n$ by

$$(c, t) = \left(bA + C, A^{-1}s \right).$$

Because $A \in GL(n, \mathbb{Z})$, and because both period lattices Λ^F and Λ^G are standard, the matrix A^{-1} sends Λ^F isomorphically to Λ^G and descends to a well-defined diffeomorphism $\Phi \colon F(X) \times T^n \to G(X) \times T^n$. We need to show Φ is symplectic. We have $dG_j = \sum_i dF_i A_{ij}$ and $dt_j = \sum_k A_{jk}^{-1} ds_k$, so

$$\sum_j dG_j \wedge dt_j = \sum_{i,j,k} A_{ij} A_{jk}^{-1} dF_i \wedge ds_k = \sum_{i,k} \delta_{ik} dF_i \wedge ds_k = \sum_i dF_i \wedge ds_i,$$

which shows that Φ is a symplectic map.

Note that, by construction,

$$\Phi(\sigma(\boldsymbol{b})) = \Phi(\boldsymbol{b}, 0) = (\boldsymbol{b}A + \boldsymbol{C}, 0) = \tau(\phi(\boldsymbol{b})).$$

If Φ' were another symplectomorphism fibred over ϕ with this property then $\Phi^{-1} \circ \Phi'$ would be a fibred automorphism of $\boldsymbol{F}$ fixing σ, and hence equal to the identity by Lemma 2.24. $\qquad\square$

From now on, we will suppose for convenience that $\phi = \mathrm{id}$, so that we have two regular Lagrangian fibrations $f \colon X \to B$ and $g \colon Y \to B$ which equip B with the same integral affine structure and we ask if there is a fibred symplectomorphism $\Phi \colon X \to Y$.

Theorem 2.26 *Suppose that we have regular Lagrangian fibrations $f \colon X \to B$ and $g \colon Y \to B$ over the same integral affine base. Suppose, moreover, that both fibrations admit global Lagrangian sections σ and τ. Then, there is a unique fibred symplectomorphism $\Phi \colon X \to Y$ such that $\Phi \circ \sigma = \tau$.*

Proof Given a sufficiently small $U \subseteq B$, Lemma 2.25 produces a unique fibred symplectomorphism $\Phi_U \colon f^{-1}(U) \to g^{-1}(U)$ satisfying $\Phi_U \circ \sigma = \tau$. We would like to define Φ by $\Phi(x) = \Phi_U(x)$ if $f(x) \in U$. The only thing to check is that this prescription is well-defined independently of the choice of U. In other words, given subsets $U, V \subseteq B$, and $x \in X$ such that $f(x) \in U \cap V$, we want to show that $\Phi_U(x) = \Phi_V(x)$. Since $\Phi_U \circ \sigma = \tau$ and $\Phi_V \circ \sigma = \tau$, we see that the restrictions of these fibred symplectomorphisms to $f^{-1}(U \cap V)$ must agree by the uniqueness part of Lemma 2.25, so $\Phi_U(x) = \Phi_V(x)$, as required. $\qquad\square$

The assumption that there is a global Lagrangian section is necessary, as the following example illustrates.

Example 2.27 Consider the quotient K of the product $\mathbb{R} \times T^3$ by the equivalence relation $(t, x, y, z) \sim (t+1, x, y, y+z)$. The symplectic form $\omega = dt \wedge dx + dy \wedge dz$ descends to K because $d(t+1) \wedge dx + dy \wedge d(y+z) = dt \wedge dx + dy \wedge dz$. The symplectic manifold (K, ω) is called the Kodaira–Thurston manifold and was the first known example of a symplectic manifold which does not admit a compatible Kähler structure;[7] see [107].

The projection $(t, x, y, z) \mapsto (t, y)$ is a well-defined regular Lagrangian fibration $K \to T^2$. The action of $(\theta_1, \theta_2) \in T^2$ by $(t, x, y, z) \mapsto (t, x + \theta_1, y, z + \theta_2)$ has the fibres of f as its orbits. If there were a section[8] $T^2 \to K$, say $(t, y) \mapsto (t, x(t, y), y, z(t, y))$, then one would get a diffeomorphism $T^4 \to K$, $(t, y, \theta_1, \theta_2) \mapsto (t, x(t, y) + \theta_1, y, z(t, y) + \theta_2)$. There is no such diffeomorphism because $K \not\cong T^4$ (for example, $b_1(K) = 3 \neq 4 = b_1(T^4)$). Therefore, there is no section.

The base of this fibration is the torus T^2 with its product integral affine structure. This same integral affine manifold arises as the base of a different Lagrangian fibration: the standard torus fibration $T^4 \to T^2$, where we equip T^4 with the symplectic form $d\theta_1 \wedge d\theta_2 + d\theta_3 \wedge d\theta_4$ and the torus fibration is $\theta \mapsto (\theta_1, \theta_3)$. This shows that it is possible to have two inequivalent Lagrangian fibrations over the same integral affine base, provided one of them does not admit a global Lagrangian section.

Remark 2.28   In fact, one can also compare two Lagrangian fibrations $f \colon X \to B$ and $g \colon Y \to B$ without assuming the existence of a global Lagrangian section. Given a subset $U \subseteq B$, consider the set $\mathcal{S}(U)$ of fibred symplectomorphisms $\Phi \colon f^{-1}(U) \to g^{-1}(U)$. This assignment $U \mapsto \mathcal{S}(U)$ is a *sheaf* over B. Using the language of sheaf theory, one can formulate an analogue of Theorem 2.26 without mentioning Lagrangian sections. There is an element $\Phi \in \mathcal{S}(B)$ (i.e. a fibred symplectomorphism) if and only if a certain characteristic class vanishes. See [26, Section 2] for a full discussion.

When we do have global Lagrangian sections, Theorem 2.26 is a wonderful compression of information: to reconstruct our $2n$-dimensional space X, all we need is an n-dimensional integral affine manifold. For example, if $n = 2, 3$, this brings four- and six-dimensional spaces into the range of visualisation.

2.6 Lagrangian and Non-Lagrangian Sections

We now turn to the question of when a Lagrangian fibration admits a Lagrangian section. First, we see what happens to the symplectic form in Liouville coordinates when we pick a non-Lagrangian section.

Lemma 2.29   *Let $H \colon X \to \mathbb{R}^n$ be an integrable Hamiltonian system, let*

---

[7] You can see this because, for example, the first Betti number of a Kähler manifold must be even, but $b_1(K) = 3$.

[8] Lagrangian or not.

$B \subseteq H(X) \subseteq \mathbb{R}^n$ *be an open set, and let* $\sigma \colon B \to X$ *be a (not necessarily Lagrangian) section. Define*

$$\Psi \colon B \times \mathbb{R}^n \to X, \qquad \Psi(\boldsymbol{b}, \boldsymbol{t}) = \phi_t^H(\sigma(\boldsymbol{b})).$$

Let β *denote the pullback of the 2-form* $\sigma^*\omega$ *on* B *to* $B \times \mathbb{R}^n$. *Then* Ψ *is both an immersion and a submersion and* $\Psi^*\omega = \sum db_i \wedge dt_i + \beta$, *where* $(b_1, \dots, b_n)$ *are the standard coordinates on* $B \subseteq \mathbb{R}^n$.

Proof The only difference with the proof of Theorem 1.38 is that the quantity $\omega(\Psi_*\partial_{b_i}, \Psi_*\partial_{b_j})$ does not need to vanish. Instead,

$$\omega(\Psi_*\partial_{b_i}, \Psi_*\partial_{b_i}) = \omega((\phi_t^H)_*\sigma_*\partial_{b_i}, (\phi_t^H)_*\sigma_*\partial_{b_j})$$
$$= \omega(\sigma_*\partial_{b_i}, \sigma_*\partial_{b_j})$$
$$= \sigma^*\omega(\partial_{b_i}, \partial_{b_j}),$$

which gives the term β in $\Psi^*\omega$ as claimed. This 2-form is still nondegenerate (each ∂_{b_i} pairs nontrivially with the corresponding ∂_{t_i}), so Ψ is still a submersion and an immersion. □

Lemma 2.30 *In the situation of the previous lemma, if there is a 1-form* η *on* B *with* $\sigma^*\omega = d\eta$ *then there is a Lagrangian section over* B.

Proof If $\tau(\boldsymbol{b}) = (\boldsymbol{b}, \boldsymbol{t}(\boldsymbol{b}))$ is another section (written with respect to the coordinate system Ψ) then we can compute $\tau^*\omega$ by following the calculation in Exercise 1.51. We get

$$\omega(\tau_*\partial_{b_i}, \tau_*\partial_{b_j}) = \frac{\partial t_i}{\partial b_j} - \frac{\partial t_j}{\partial b_i} + \beta(\partial_{b_i}, \partial_{b_j}).$$

By comparing with the formula for the exterior derivative of the 1-form $\sum t_i(\boldsymbol{b})db_i$, we see that $\tau^*\omega = d(\sum t_i(\boldsymbol{b})db_i) + \beta$. Now suppose that $\beta = d\eta$ for some 1-form $\eta = \sum \eta_i(\boldsymbol{b})db_i$. Taking $t_i(\boldsymbol{b}) = -\eta_i(\boldsymbol{b})$, we get a section for which $\tau^*\omega = -\beta + \beta = 0$, that is, a Lagrangian section. □

Corollary 2.31 *If* $H \colon X \to \mathbb{R}^n$ *is an integrable Hamiltonian system with* $H(X) = B$ *and* σ *is a section over* B *with[9]* $[\sigma^*\omega] = 0 \in H^2_{dR}(B)$ *then* H *admits a Lagrangian section over* B. *In fact, if* σ *is Lagrangian over a subset* $B' \subseteq B$ *and[10]* $[\sigma^*\omega] = 0 \in H^2_{dR}(B, B')$ *then* H *admits a Lagrangian section which agrees with* σ *over* B'.

---

[9] $H^2_{dR}(B)$ denotes the De Rham cohomology group of closed 2-forms modulo exact 2-forms; $H^2_{dR}(B) = 0$ is a fancy way of saying 'if $d\beta = 0$ then $\beta = d\eta$'.

[10] $H^2_{dR}(B, B')$ denotes the relative De Rham cohomology. This is again closed forms modulo exact forms, but where the forms β and η are required to vanish on B'. This is a slightly different formulation to the standard set-up in, say, the book by Bott and Tu [10, pp. 78–79] but equivalent to it (as explained in the MathOverflow answer [27] by Ebert).

Proof Note that $\beta := \sigma^*\omega$ is closed, so it defines a de Rham cohomology class. If $[\beta] = 0$ in de Rham cohomology then there exists a 1-form such that $\beta = d\eta$. If $\beta = 0$ on B' then it defines a class in relative de Rham cohomology $H^2_{dR}(B, B')$, which vanishes if and only if $\beta = d\eta$ for a 1-form η which itself vanishes on B'. Inspecting the proof of Lemma 2.29, this means that the Lagrangian section built using η coincides with σ on B'. □

Remark 2.32 We will use the condition on relative cohomology to find Lagrangian sections for non-regular Lagrangian fibrations: we will first construct Lagrangian sections near the critical fibres, then extend them over the regular locus using this result, provided that the relevant relative cohomology group vanishes.

Corollary 2.33 *Let $f : X \to B$ be a regular Lagrangian fibration and suppose σ is a section which is Lagrangian over a (possibly empty) subset $B' \subseteq B$. If $[\sigma^*\omega] = 0 \in H^2_{dR}(B, B')$ then f admits a Lagrangian section.*

Proof By the cohomological assumption, there exists a 1-form η on B such that $\eta = 0$ on B' and $d\eta = \sigma^*\omega$. Cover B by integral affine coordinate charts; the Lagrangian fibration is equivalent to an integrable Hamiltonian system over each of these charts, and we can apply Lemma 2.30 (using η) to modify σ and obtain a Lagrangian section. Since we are using the same 1-form on different charts, we modify σ in the same way on overlaps between charts, so we find a Lagrangian section over the whole of B. □

Remark 2.34 (Exercise 2.37) We will later apply this when B is a punctured surface and B' is a neighbourhood of a strict subset of the punctures. This satisfies $H^2(B, B') = 0$.

2.7 Solutions to Inline Exercises

Exercise 2.35 (Corollary 2.8) *If $f : X \to B$ is a proper submersion with connected Lagrangian fibres then the fibres are Lagrangian tori.*

Solution By Lemma 2.6, if we pick local coordinates $(b_1, \ldots, b_n)$ on B then the functions $b_1 \circ f, \ldots, b_n \circ f$ form an integrable Hamiltonian system, so this follows from the little Arnold–Liouville theorem (Theorem 1.40). □

Exercise 2.36 *If ω is not an exact 2-form, how can we construct the integral affine structure on B?*

Solution We need to define the flux map $\tilde{B} \to \mathbb{R}^n$. As before, we fix the universal cover $p\colon \tilde{B} \to B$ and write $\xi \to B$ for the local system with fibre $H_1(f^{-1}(b); \mathbb{Z})$ over $b \in B$. We pick a $\mathbb{Z}$-basis of global sections $c_1, \ldots, c_n$ of $\tilde{\xi} = p^*\xi$. Write $\tilde{f}\colon p^*X \to \tilde{B}$ for the pullback of f to the universal cover (i.e. the Lagrangian fibration whose fibre over $\tilde{b}$ is $f^{-1}(p(\tilde{b}))$). We continue to write ω for the pullback of ω to p^*X.

Fix a basepoint $\tilde{b}_0 \in \tilde{B}$. Given a point $\tilde{b} \in \tilde{B}$, pick a path $\gamma\colon [0, 1] \to \tilde{B}$ from $\tilde{b}_0$ to $\tilde{b}$. A *family of loops over γ* (see Figure 2.1) is a homotopy $C\colon S^1 \times [0, 1] \to p^*X$ satisfying $\tilde{f}(C(s, t)) = \gamma(t)$. That is, if t is fixed, then $C(s, t)$ is a loop in $\tilde{f}^{-1}(\gamma(t))$.

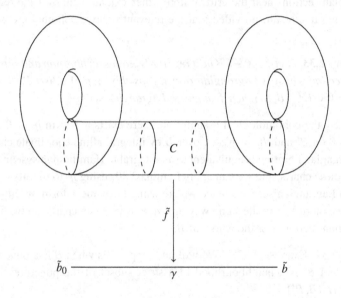

Figure 2.1 A family of loops over γ.

For $k = 1, \ldots, n$, pick a family of loops C_k over γ with $C_k(\cdot, t) \in c_k(\gamma(t))$ for all $t \in [0, 1]$. Define

$$I(\tilde{b}) = (I_1(\tilde{b}), \ldots, I_n(\tilde{b})), \qquad I_k(\tilde{b}) = \int_{C_k} \omega.$$

It remains to understand how this flux map depends on the choices we made, namely:

1 a basis $c_1, \ldots, c_n$ of $p^*\xi$,
2 a basepoint $\tilde{b}_0$,
3 a path γ from $\tilde{b}_0$ to $\tilde{b}$,
4 a family of loops C_k over γ for each $k \in \{1, \ldots, n\}$.

We deal first with the choice of γ and C_k. Since $\tilde{B}$ is simply-connected, a different choice of path γ' from $\tilde{b}_0$ to $\tilde{b}$ will be homotopic to γ via some homotopy $h\colon [0,1] \times [0,1] \to \tilde{B}$. Choose C_k over γ and C_k' over γ'. We will show that $\int_{C_k} \omega = \int_{C_k'} \omega$.

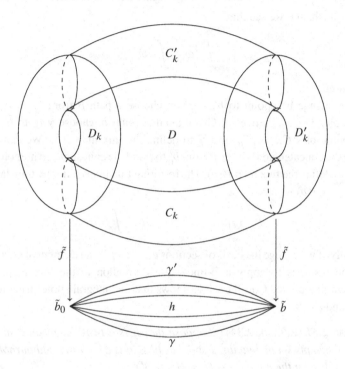

Figure 2.2 Different choices of paths and homotopies for the solution of Exercise 2.36.

The loops $C_k(\cdot,0)$ and $C_k'(\cdot,0)$ are homologous in $\tilde{f}^{-1}(\gamma(0))$ by assumption, and therefore freely homotopic because $\pi_1(T^n) \cong H_1(T^n;\mathbb{Z}) \cong \mathbb{Z}^n$. Let $D_k\colon S^1 \times [0,1] \to \tilde{f}^{-1}(\gamma(0))$ be a free homotopy with $D_k(\cdot,0) = C_k(\cdot,0)$ and $D_k(\cdot,1) = C_k'(\cdot,0)$. By the homotopy lifting property of the submersion $\tilde{f}$, we can find a map $D\colon S^1 \times [0,1] \times [0,1] \to p^*X$ with $\tilde{f} \circ D = h$ and $D(s,t,0) = D_k(s,t)$. Define $D_k'(s,t) = D(s,t,1)$; this defines a cylinder in the Lagrangian torus $\tilde{f}^{-1}(\gamma(1))$ (see Figure 2.2). Consider D as a 3-chain (in the sense of singular homology). Because $d\omega = 0$, we have

$$0 = \int_D d\omega = \int_{\partial D} \omega$$

by Stokes's theorem. But $\partial D = C_k + D'_k - C'_k - D_k$, so

$$0 = \int_{C_k} \omega + \int_{D'_k} \omega - \int_{C'_k} \omega - \int_{D_k} \omega.$$

Since D_k and D'_k are contained in Lagrangian fibres, the integrals $\int_{D_k} \omega$ and $\int_{D'_k} \omega$ vanish, and we see that

$$\int_{C_k} \omega = \int_{C'_k} \omega$$

as required.

If we change basepoint to $\tilde{b}'_0$, we can choose a path β from $\tilde{b}'_0$ to $\tilde{b}_0$ and homotopies $\Gamma_1, \ldots, \Gamma_n$ over β. Given another point $\tilde{b}$, choose γ from $\tilde{b}_0$ to $\tilde{b}$ and homotopies $C_1, \ldots, C_n$ over γ to define the flux map $I(\tilde{b})$. We can then choose the concatenated path $\gamma \cdot \beta$ from $\tilde{b}'_0$ to $\tilde{b}$ and the concatenated homotopies $C_k \cdot \Gamma_k$ to define the flux map $I'(\tilde{b})$. The resulting flux maps differ by translation: $I'(\tilde{b}) = J + I(\tilde{b})$ with

$$J = (J_1, \ldots, J_n), \qquad J_k = \int_{\Gamma_k} \omega.$$

Finally, if we change the basis of sections $c_1, \ldots, c_n$ by an element of $GL(n, \mathbb{Z})$ then the result is to apply a $\mathbb{Z}$-linear transformation to the flux map. The argument that proved Corollary 2.18 shows that the integral affine structure on $\tilde{B}$ descends to B. □

Exercise 2.37 (Remark 2.34) *Suppose that B is a two-dimensional surface with a nonempty set of punctures and that $B' \subseteq B$ is a collar neighbourhood of a strict subset of the punctures. Then $H^2(B, B') = 0$.*

Proof Note first that the second cohomology of a punctured surface is zero (provided there is at least one puncture). We have $H^2(B, B') \cong H^2(B/B')$. The quotient B/B' is the result of filling in a strict subset of the punctures, so it is homeomorphic to a surface with fewer (but still some) punctures. Therefore $H^2(B, B') = H^2(B/B') = 0$. □

3

Global Action-Angle Coordinates and Torus Actions

3.1 Hamiltonian Torus Actions

One way of stating the Arnold–Liouville theorem is that, after a suitable change of coordinates in the target, the $\mathbb{R}^n$-action generated by the Hamiltonian vector fields $V_{H_1}, \ldots, V_{H_n}$ actually factors through a T^n-action. In this chapter, we work backwards, assuming that we have a globally defined torus action, even on the non-regular fibres, and see what kinds of critical points can occur.

Definition 3.1 Let $H \colon X \to \mathbb{R}^n$ be an integrable Hamiltonian system such that the Hamiltonian $\mathbb{R}^n$-action ϕ_t^H factors through a Hamiltonian T^n-action, that is, $\phi_t^H = \mathrm{id}$ for any $t \in (2\pi\mathbb{Z})^n$. Then we call H the *moment map* for the torus action. It is conventional to write μ rather than H for a moment map, and we will do this wherever we want to emphasise the existence of the torus action. We will call a symplectic $2n$-manifold X a *toric manifold* if it admits a Hamiltonian T^n-action.

We saw in Lemma 2.25 that the image of a moment map determines the Hamiltonian system completely up to fibred symplectomorphism, at least if there are no critical points and there is a global Lagrangian section. We therefore concentrate on the image $\mu(X)$ of the moment map, which we will call the *moment image* or *moment polytope*. The Atiyah–Guillemin–Sternberg convexity theorem, discussed in Section 3.2 shortly, tells us that $\mu(X)$ is indeed a rational convex polytope. We will not give a full proof of this theorem, as there are many excellent expositions in the literature (e.g. Atiyah [4, Theorem 1], Audin [5], Guillemin and Sternberg [49, Theorem 4], McDuff and Salamon [77, Theorem 5.47], amongst others). Instead, we will prove the much easier Proposition 3.3: that under mild conditions, the boundary of the moment image is piecewise linear. This has the advantage of being a local result, which will apply in situations where we only have a torus action on some parts of the

31

manifold. In particular, it will apply in situations where there is no sense in which the image of the Hamiltonian system is convex, like the almost toric setting in Chapter 8.

We need the following preliminary lemma:

Lemma 3.2 (Exercise 3.24) *Let $\mu\colon X \to \mathbb{R}^n$ be the moment map of a Hamiltonian T^n-action. If $s\colon \mathbb{R}^n \to \mathbb{R}$ is a linear map, $s(b_1, \ldots, b_n) = \sum s_i b_i$ then $s \circ \mu$ generates the Hamiltonian flow $\phi^\mu_{(s_1 t, \ldots, s_n t)}$.*

In this case, the $\mathbb{R}$-action (flow) $\phi^{s \circ \mu}_t$ can be thought of as a subgroup of the T^n action, coming from the homomorphism

$$s^T\colon \mathbb{R} \to \mathbb{R}^n / (2\pi\mathbb{Z})^n, \qquad s^T(t) = (s_1 t, \ldots, s_n t).$$

Now suppose that X is a symplectic $2n$-manifold, and that $\mu\colon X \to \mathbb{R}^n$ is the moment map for a Hamiltonian T^n-action. Write $\partial\mu(X)$ for the boundary of the moment image. We will assume that $\partial\mu(X)$ is a piecewise smooth hypersurface; we will show that, under mild assumptions, $\partial\mu(X)$ is piecewise *linear*. Pick local smooth embeddings $\delta_i\colon (0,1)^{n-1} \to \partial\mu(X) \subseteq \mathbb{R}^n$ parametrising the smooth pieces of $\partial\mu(X)$ and assume that there are smooth lifts $\gamma_i\colon (0,1)^{n-1} \to X$ such that $\delta_i = \mu \circ \gamma_i$.

Proposition 3.3 (Piecewise linearity of the toric boundary) *The image of each δ_i is contained in an affine hyperplane Π_i with rational slopes, that is, $\Pi_i = \{x \in \mathbb{R}^n : \alpha \cdot x = c\}$ for some integer vector α. If $z \in \mu^{-1}(\delta_i)$ then the stabiliser of z is precisely the one-dimensional subtorus $s^T_i(\mathbb{R}) \subseteq T^n$ where $s_i(x) = \alpha \cdot x$.*

Proof Let $\Pi_i(t)$ be the tangent hyperplane to δ_i at $\delta_i(t)$, with normal vector $\alpha = (\alpha_1, \ldots, \alpha_n)$. We say that δ_i has rational slopes at b if α is parallel to an integer vector. Otherwise, at least one of the ratios α_k / α_ℓ is irrational. We will show that the tangent hyperplane to $\Pi_i(t)$ has rational slopes for all t, which is only possible if $\Pi_i(t)$ is independent of t (otherwise, the slopes would need to take irrational values by the intermediate value theorem). This will imply that δ_i coincides with its tangent hyperplane. The statement about stabilisers will come up naturally in the proof.

Suppose that δ_i has an irrational slope at $b := \delta_i(t)$. Pick a ball B centred at b and a smooth function $S\colon B \to \mathbb{R}$ such that $\mu(X) \cap B = \{p \in B : S(p) \geq 0\}$ and $\partial\mu(X) \cap B = S^{-1}(0)$ is a regular level set. The function $S \circ \mu$ has a minimum along $S^{-1}(0)$, so if $z \in \mu^{-1}(b)$ then $d_b S \circ d_z \mu = 0$. Let $s := d_b S$ and consider the Hamiltonian function $H := s \circ \mu$; by Lemma 3.2, this generates the $\mathbb{R}$-action given by $s^T(\mathbb{R}) \subseteq T^n$. But $d_z H = s \circ d_z \mu = 0$, so this $\mathbb{R}$-action fixes any point $z \in \mu^{-1}(b)$. The stabiliser of z is a closed subgroup of T^n containing

$s^T(\mathbb{R}) \subseteq T^n$; if δ_i has an irrational slope at b then the closure of this subgroup is at least two-dimensional, so the stabiliser of z contains a 2-torus. This means there are two linearly independent components of μ whose Hamiltonian vector fields vanish at z; in particular, the rank of $d_{\gamma_i(t)}\mu$ is at most $n - 2$. Since $\delta_i = \mu \circ \gamma_i$, we have $d_t\delta_i = d_{\gamma_i(t)}\mu \circ d_t\gamma_i$, and this means that the rank of $d_t\delta_i$ is at most $n - 2$. This contradicts the assumption that δ_i is an embedding (δ_i fails to be an immersion at t).

If δ_i has rational slopes then we can take $S(x) = \alpha \cdot x$ as the function which is constant along δ_i, and the same argument gives us the stabiliser as claimed. $\square$

Remark 3.4 As this book progresses, we will allow our Lagrangian torus fibrations $f : X \to B$ to have more and more different types of critical points. If $B^{reg} \subseteq B$ denotes the set of regular values of f then we know B^{reg} inherits an integral affine structure. We can now allow f to have 'toric critical points', where X admits a local Hamiltonian torus action having f as its moment map. Proposition 3.3 tells us that B will have the structure of an integral affine manifold with piecewise linear boundary and corners, extending the integral affine structure on B^{reg}.

3.2 Delzant Polytopes and Toric Manifolds

Definition 3.5 A *rational convex polytope* P is a subset of $\mathbb{R}^n$ defined as the intersection of a finite collection of half-spaces $S_{\alpha,b} = \{x \in \mathbb{R}^n : \alpha_1 x_1 + \cdots + \alpha_n x_n \le b\}$ with $\alpha_1, \ldots, \alpha_n \in \mathbb{Z}$ and $b \in \mathbb{R}^n$. We say that P is a *Delzant*[1] *polytope* if it is a convex rational polytope such that every point on a k-dimensional facet has a neighbourhood isomorphic (via an integral affine transformation) to a neighbourhood of the origin in the polytope $[0, \infty)^{n-k} \times \mathbb{R}^k$. A vertex of a polytope is called *Delzant* if the germ of the polytope at that vertex is Delzant.

Example 3.6 The polygon in Figure 3.1 fails to be Delzant: there is no integral affine transformation sending the marked vertex to the origin and sending the two marked edges to the x- and y-axes. Indeed, the primitive integer vectors $(0, 1)$ and $(2, 1)$ pointing along these edges span a strict sublattice of the integer lattice $\mathbb{Z}^2$.

Theorem 3.7 *Let X be a toric manifold, that is, a symplectic $2n$-manifold equipped with a Hamiltonian T^n-action with moment map $\mu : X \to \mathbb{R}^n$.*

1 (Atiyah–Guillemin–Sternberg convexity theorem [4, Theorem 1], [49, Theorem 4]) The moment image $\Delta := \mu(X)$ is a Delzant polytope. If X is compact,

---

[1] Audin [5] calls these *primitive polytopes*.

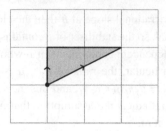

Figure 3.1 A non-Delzant polygon.

then Δ is the convex hull of $\{\mu(x) \; : \; x \in Fix(X)\}$, where $Fix(X)$ is the set of fixed points of the torus action.

2 *(Delzant existence theorem [24, Section 3]) For any compact Delzant polytope $\Delta \subseteq \mathbb{R}^n$ there exists a symplectic 2n-manifold X_Δ and a map $\mu \colon X_\Delta \to \mathbb{R}^n$ with $\mu(X_\Delta) = \Delta$ such that μ generates a Hamiltonian T^n-action. Moreover, X_Δ is a projective variety. Such varieties are often called* projective toric varieties.

3 *(Delzant uniqueness theorem [24, Theorem 2.1]) The moment polytope determines the pair (X, μ) up to fibred symplectomorphism.*

We will not prove (1) or (3). We will see two constructions of X_Δ later, proving (2). In the remainder of this chapter, we will focus instead on examples where we can extract geometric information about X from the moment polytope.

3.3 Examples

Example 3.8   Consider the n-torus action on $\mathbb{C}^n$ given by

$$(z_1, \ldots, z_n) \mapsto (e^{it_1} z_1, \ldots, e^{it_n} z_n).$$

This is Hamiltonian, with moment map

$$\mu(z_1, \ldots, z_n) = \left(\frac{1}{2}|z_1|^2, \ldots, \frac{1}{2}|z_n|^2 \right).$$

The image of the moment map is the nonnegative orthant. This is a manifold with boundary and corners: the μ-preimage of a boundary stratum of codimension k is an $(n - k)$-dimensional torus. For example, the preimage of the vertex is a single fixed point (the origin), the preimage of a point on the positive b_1-axis is a circle with fixed radius in the z_1-plane, the preimage of a point on the interior of the $b_1 b_2$-plane is a 2-torus, and so forth.

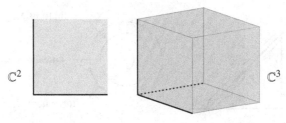

$\mathbb{C}^2$ $\mathbb{C}^3$

Remark 3.9   The critical values of μ are precisely the boundary points of the moment polytope. The boundary is stratified into facets of dimension 0 (vertices), 1 (edges), 2 (faces), and so on, so we can classify the critical values according to the dimension of the stratum to which they belong. By definition, any Delzant polytope is locally isomorphic to $\mathbb{R}^k \times [0, \infty)^{n-k}$ in a neighbourhood of a point in a k-dimensional facet. In Example 3.8, we have found a system whose moment image is $[0, \infty)^{n-k}$, so by Theorem 3.7(3), this means that the integrable Hamiltonian system in a neighbourhood of a critical point living over a k-dimensional facet is fibred-symplectomorphic to the system

$$\mu \colon \mathbb{R}^k \times (S^1)^k \times \mathbb{C}^{n-k} \to \mathbb{R}^n,$$

$$\mu(p, q, z_{k+1}, \ldots, z_n) = \left(p, \frac{1}{2}|z_{k+1}|^2, \ldots, \frac{1}{2}|z_n|^2 \right).$$

Such critical points are called *toric*,[2] and the set of all toric critical points is often called the *toric boundary* of X. It is not a boundary in the usual sense: it is a union of submanifolds of codimension 2. Instead, considering X as a projective variety, it is the boundary in the sense of algebraic geometry: it is a divisor, and is often called the *toric divisor*.

Here is a nice way to understand the genus 1 Heegaard decomposition of the 3-sphere using the moment map for $\mathbb{C}^2$.

Example 3.10 (Heegaard decomposition of S^3)   Let $\mu \colon \mathbb{C}^2 \to \mathbb{R}^2$ be the moment map from Example 3.8. The preimage of the line segment $b_1 + b_2 = \frac{1}{2}$, $b_1, b_2 \geq 0$, is the subset $S := \{(z_1, z_2) \in \mathbb{C}^2 : |z_1|^2 + |z_2|^2 = 1\}$, that is, the unit 3-sphere; this is the slanted line segment in Figure 3.2. The fibre $T := \mu^{-1}\left(\frac{1}{4}, \frac{1}{4} \right)$ is a torus with $T \subseteq S$. We can see from Figure 3.2 that T separates S into two pieces S_1, S_2, and it is also easy to see that each piece is homeomorphic to a *solid torus* $S^1 \times D^2$: the 'core circles' of these solid tori are the fibres $s_1 = \mu^{-1}\left(\frac{1}{2}, 0 \right)$, $s_2 = \mu^{-1}\left(0, \frac{1}{2} \right)$ over the points where the line segment intersects the b_1- and b_2-axes.

---

[2]   In fact, it is a theorem of Eliasson [30] and Dufour and Molino [25] that toric critical points can be characterised purely in terms of the Hessian of the Hamiltonian system at the critical point. They call such critical points *elliptic*.

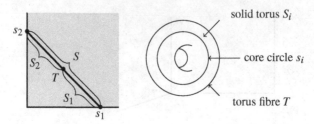

Figure 3.2 The unit sphere in $\mathbb{C}^2$ lives over the slanted line; the fibre T separates it into two solid tori.

Example 3.11 (Exercise 3.25)   Consider the unit 2-sphere (S^2, ω) where ω is the area form. By comparing infinitesimal area elements, one can show that the projection map from S^2 to a circumscribed cylinder is area-preserving.[3] Let $\mu \colon S^2 \to \mathbb{R}$ be the height function $\mu(x, y, z) = z$ (thinking of S^2 embedded in the standard way in $\mathbb{R}^3$). Then μ is a moment map for the circle action which rotates around the z-axis. The moment image is $[-1, 1] \subseteq \mathbb{R}$.

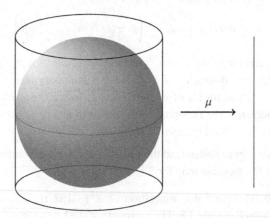

Example 3.12   If we take S^2 with the area form $\lambda\omega$ (where ω is the form giving area 4π) then the rescaled height function λz is a moment map for the circle action which rotates around the z-axis with period 2π. The moment image is $[-\lambda, \lambda]$.

Example 3.13   One can form more examples by taking products. If we take $S^2 \times \cdots \times S^2$ with the product symplectic form giving the ith factor symplectic area $4\pi\lambda_i$ then we get a T^n-action on $(S^2)^n$, whose moment map

---

[3] If Cicero is to be believed [20, XXIII–64,65], a diagram representing this theorem was engraved on the tomb of Archimedes (who proved it).

is $\mu((x_1, y_1, z_1), \ldots, (x_n, y_n, z_n)) = (z_1, \ldots, z_n)$, with image the hypercuboid $[-\lambda_1, \lambda_1] \times \cdots \times [-\lambda_n, \lambda_n]$. For example, if we use equal areas $\lambda_1 = \cdots = \lambda_n = 1$ then the moment image for $S^2 \times S^2$ is a square, whose vertices correspond to the fixed points $\{(0, 0, \pm 1)\} \times \{(0, 0, \pm 1)\}$, and whose edges correspond to the spheres $S^2 \times \{(0, 0, \pm 1)\}$ and $\{(0, 0, \pm 1)\} \times S^2$. For $S^2 \times S^2 \times S^2$ the moment image is a cube whose horizontal faces correspond to the submanifolds $S^2 \times S^2 \times \{(0, 0, \pm 1)\}$, and so on.

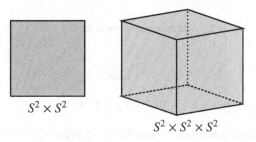

$$S^2 \times S^2 \qquad S^2 \times S^2 \times S^2$$

Definition 3.14 (Affine length) If $\ell \colon [0, L] \to \mathbb{R}^n$ is a line segment of the form $\ell(t) = at + b$ with $a \in \mathbb{Z}^n$ a primitive vector[4] and $b \in \mathbb{R}^n$ then we say ℓ is a *rational* line segment and define the *affine length* of ℓ to be L.

Example 3.15 Consider the triangle in Figure 3.1. The horizontal edge has affine length 2 and the other two edges both have affine length 1.

Lemma 3.16 *If $\ell \colon [0, L] \to \mathbb{R}^n$ is a rational line segment whose image is an edge of the moment polytope then $\mu^{-1}(\ell([0, L]))$ is a symplectic sphere of symplectic area $2\pi L$.*

Proof By Theorem 3.7(3), the preimage of an edge is determined up to fibred symplectomorphism by its moment image $\ell([0, L])$. By comparing with Example 3.12, we see that the preimage of such an edge is symplectomorphic to $\left(S^2, \frac{L\omega}{2} \right)$. □

Example 3.17 Consider the complex projective n-space $\mathbb{CP}^n$, with homogeneous coordinates $[z_1 : \cdots : z_{n+1}]$ (see Appendix C). This has a torus action $[z_1 : \cdots : z_{n+1}] \mapsto [e^{it_1} z_1 : \cdots : e^{it_n} z_n : z_{n+1}]$ which is Hamiltonian, for the Fubini–Study form[5] ω, with moment map

$$\mu([z_1 : \cdots : z_{n+1}]) = \left(\frac{|z_1|^2}{|z|^2}, \ldots, \frac{|z_n|^2}{|z|^2} \right),$$

---

[4] An integer vector a is called primitive if it is a shortest integer vector on the line it spans, in other words, if $\lambda a \in \mathbb{Z}^n$ implies $|\lambda| \geq 1$.

[5] If you are not familiar with this symplectic form, we will construct it in Example 4.9.

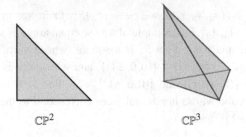

Figure 3.3 The moment polytopes for $\mathbb{CP}^2$ and $\mathbb{CP}^3$.

where $|z|^2 = \sum_{i=1}^{n+1} |z_i|^2$. The moment image is the simplex

$$\{(b_1, \ldots, b_n) \in \mathbb{R}^n \ : \ b_1, \ldots, b_n \geq 0, \ b_1 + \cdots + b_n \leq 1\}.$$

For example, $\mu(\mathbb{CP}^2)$ and $\mu(\mathbb{CP}^3)$ are drawn in Figure 3.3. In each case, the *hyperplane at infinity* $\{[z_1 : \cdots : z_n : 0]\}$ projects via μ to the facet $b_1 + \cdots + b_n = 1$ of the simplex.

Example 3.18 The *tautological bundle* over $\mathbb{CP}^1$ is the variety

$$O(-1) := \{(z_1, z_2, [z_3 : z_4]) \in \mathbb{C}^2 \times \mathbb{CP}^1 \ : \ z_1 z_4 = z_2 z_3\}.$$

This has a holomorphic projection $\pi \colon O(-1) \to \mathbb{CP}^1$, $\pi(z_1, z_2, [z_3 : z_4]) = [z_3 : z_4]$, which exhibits it as the total space of a holomorphic line bundle over $\mathbb{CP}^1$. This is a fancy way of saying that $\pi^{-1}([z_3 : z_4])$ is a complex line (specifically $\{(z_1, z_2) \in \mathbb{C}^2 \ : \ z_1 z_4 = z_2 z_3\} \subseteq \mathbb{C}^2$) for all $[z_3 : z_4] \in \mathbb{CP}^1$. The symplectic form $\omega_{\mathbb{C}^2} \oplus \omega_{\mathbb{CP}^1}$ on $\mathbb{C}^2 \times \mathbb{CP}^1$ pulls back to a symplectic form on $O(-1)$, with respect to which the following T^2-action is Hamiltonian:

$$(z_1, z_2, [z_3 : z_4]) \mapsto (e^{it_1} z_1, e^{it_2} z_2, [e^{it_1} z_3 : e^{it_2} z_4]).$$

The moment map is the sum of the moment maps for $\mathbb{C}^2$ and $\mathbb{CP}^1$:

$$\mu(z_1, z_2, [z_3 : z_4]) = \left(\frac{1}{2}|z_1|^2 + \frac{|z_3|^2}{|z_3|^2 + |z_4|^2}, \ \frac{1}{2}|z_2|^2 + \frac{|z_4|^2}{|z_3|^2 + |z_4|^2}\right).$$

The image of the moment map is the subset in Figure 3.4:

$$\Delta_{O(-1)} := \{(b_1, b_2) \in \mathbb{R}^2 \ : \ b_1, b_2 \geq 0, \ b_1 + b_2 \geq 1\}.$$

The zero-section $\mathbb{CP}^1 = \{z_1 = z_2 = 0\} \subseteq O(-1)$ projects down to the edge $b_1 + b_2 = 1$. An alternative moment map can be obtained by postcomposing

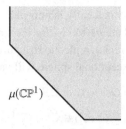

Figure 3.4 The moment polygon $\Delta_{O(-1)}$.

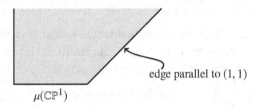

edge parallel to $(1, 1)$

$\mu(\mathbb{CP}^1)$

Figure 3.5 Alternative moment polygon for $O(-1)$.

with the integral affine transformation[6]

$$(b_1, b_2) \mapsto (b_1, b_2) \begin{pmatrix} 1 & 1 \\ 0 & 1 \end{pmatrix} + (0, -1),$$

which sends the moment polygon to

$$\{(b_1, b_2) \in \mathbb{R}^2 \; : \; b_1, b_2 \geq 0, \; b_1 - b_2 \leq 1\}.$$

This is an important example because of the role played by $O(-1)$ in birational geometry. The projection $\varpi \colon O(-1) \to \mathbb{C}^2$ given by $\varpi(z_1, z_2, [z_3 : z_4]) = (z_1, z_2)$ is a birational map called the *blow-down* or *contraction* of a -1-curve. It is an isomorphism away from $(0, 0) \in \mathbb{C}^2$, but it contracts the sphere $\{(0, 0, [z_3 : z_4]) \; : \; [z_3 : z_4] \in \mathbb{CP}^1\}$ (known as the *exceptional sphere*) to the origin.

When we introduce the symplectic cut operation in Section 4.3, we will see that if we take a toric variety X_Δ and blow up a fixed point of the torus action (living over a vertex $v \in \Delta$), we get a new toric variety $X_{\Delta'}$ whose moment polytope Δ' differs from the previous one by truncating at the vertex v. More precisely, we use an integral affine transformation to put Δ in such a position

---

[6] This is the first instance of the notation mentioned in the preface: the angle bracket reminds the reader that our matrix acts from the right. This will be more important when the matrix appears in isolation.

that v sits at the origin and Δ is locally isomorphic to $[0, \infty)^n$ near v, then we truncate Δ using the hyperplane $b_1 + \cdots + b_n = c$ for some positive c. Varying the constant c will give different symplectic structures (in particular, for $n = 2$, the symplectic area of the exceptional sphere will vary).

Example 3.19   The bundle $O(-n)$ over $\mathbb{CP}^1$ is the variety[7]

$$O(-n) := \{(z_1, z_2, [z_3 : z_4]) \in \mathbb{C}^2 \times \mathbb{CP}^1 \; : \; z_1 z_4^n = z_2 z_3^n\}.$$

The Hamiltonians

$$H_1 = \frac{1}{2}|z_1|^2 + \frac{|z_3|^2}{|z_3|^2 + |z_4|^2}, \qquad H_2 = \frac{1}{2}|z_2|^2 + \frac{|z_4|^2}{|z_3|^2 + |z_4|^2}$$

still generate circle actions, but the period lattice for the $\mathbb{R}^2$-action generated by (H_1, H_2), while constant, is no longer standard: the element $\phi_{2\pi/n}^{H_1}\phi_{2\pi/n}^{H_2}$ now acts as the identity. This means that the period lattice is spanned by $\mathbb{Z}\begin{pmatrix} 2\pi/n \\ 2\pi/n \end{pmatrix} \oplus \mathbb{Z}\begin{pmatrix} 2\pi \\ 0 \end{pmatrix}$. If we use the combination $\mu = \left(H_1, \frac{H_1+H_2}{n}\right)$ then we get a standard period lattice, so this is a valid moment map. This has the effect of applying the affine transformation $\begin{pmatrix} 1 & 1/n \\ 0 & 1/n \end{pmatrix}$ to the moment polygon in Figure 3.4; we also translate by $(0, -1/n)$ so that the horizontal edge $\mu(\mathbb{CP}^1)$ sits on the b_1-axis.

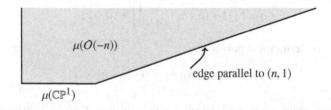

$\mu(O(-n))$

edge parallel to $(n, 1)$

$\mu(\mathbb{CP}^1)$

Figure 3.6 The moment polygon for $O(-n)$.

Similarly, one can define the bundles $O(n) \to \mathbb{CP}^1$, $n \geq 0$, and these admit torus actions; the moment map now sends a neighbourhood of the zero-section in $O(n)$ to the region shown in Figure 3.7. For example, a complex line in $\mathbb{CP}^2$ has normal bundle $O(1)$, and in the moment image of $\mathbb{CP}^2$ we see precisely the $n = 1$ neighbourhood surrounding the b_1-axis.

The following lemma now follows immediately from these examples and Theorem 3.7(3).

---

[7] The discerning reader will spot that this is the pullback of $O(-1)$ along the degree n holomorphic map $\mathbb{CP}^1 \to \mathbb{CP}^1$, $[z_3 : z_4] \mapsto [z_3^n : z_4^n]$.

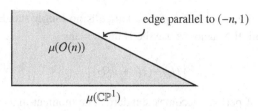

Figure 3.7 $O(n)$ for $n \geq 0$.

Lemma 3.20 *Let* $\Delta \subseteq \mathbb{R}^2$ *be a moment polygon and* $e \subseteq \Delta$ *an edge connecting two vertices* P, Q. *Assume that this edge is traversed from* P *to* Q *as you move anticlockwise around the boundary of* Δ. *Let* v, w *be primitive integer vectors pointing along the other edges emerging from* P *and* Q *respectively. Then a neighbourhood of* $\mu^{-1}(e)$ *in* X_Δ *is symplectomorphic to a neighbourhood of the zero-section in* $O(n)$ *where* $n = \det M$ *where* M *is the matrix with rows* v, w *(you may also see* $\det M$ *written as* $v \wedge w$*)*.

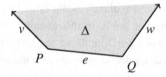

Proof　This is easily checked for the local models discussed earlier, and any edge is integral affine equivalent to one of these local models. It is therefore enough to check that $v \wedge w$ is unchanged by an integral affine transformation. The determinant is unchanged by orientation-preserving integral affine transformations. An orientation-reversing transformation will switch the sign of $v \wedge w$, but also switch the order to $w \wedge v$ because it switches anticlockwise to clockwise, so these sign effects will cancel. □

3.4 Non-Delzant Polytopes

Example 3.21　Consider the group of nth roots of unity μ_n acting on $\mathbb{C}^2$ via $(z_1, z_2) \mapsto (\mu z_1, \mu^a z_2)$ where $\gcd(a, n) = 1$. Let $X = \mathbb{C}^2/\mu_n$ be the quotient by this group action. This is a symplectic *orbifold*: the origin is a singular point. We call this kind of singularity a *cyclic quotient singularity of type* $\frac{1}{n}(1, a)$.

Hamiltonian flows still make perfect sense on X provided they fix the origin. Consider the Hamiltonians $H_1 = \frac{1}{2}|z_1|^2$ and $H_2 = \frac{1}{2}|z_2|^2$; these are invariant under the action of μ_n and hence define functions on X. The flow is simply $(e^{it_1}z_1, e^{it_2}z_2)$. However, the period lattice is no longer standard; we have $\phi_{H_1}^{2\pi/n}\phi_{H_2}^{2\pi a/n} = \text{id}$. If instead we use the Hamiltonians

$$\left(H_2, \frac{1}{n}(H_1 + aH_2)\right)$$

then the lattice of periods becomes standard. The moment image is a convex wedge in the plane bounded by the rays emanating from the origin in directions $(0, 1)$ and (n, a); we will denote this noncompact polygon by $\pi(n, a)$:

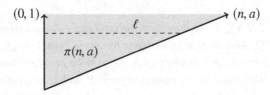

This polygon is not Delzant at the origin, corresponding to the fact that X is not smooth at the origin.

Remark 3.22   The *link* of a singularity is the boundary of a small Euclidean neighbourhood of the singular point. In this example, the link of the $\frac{1}{n}(1, a)$-singularity is the preimage of a horizontal line segment ℓ running across $\pi(n, a)$. As in Example 3.10, this has a decomposition as a union of two solid tori; this means it is a *lens space*. By definition, this is the lens space $L(n, a)$.

Lemma 3.23 (Exercise 3.26)   *The lens space $L(n, a + kn)$ is diffeomorphic to $L(n, a)$ for all integers k. The lens space $L(n, a)$ is diffeomorphic to $L(n, \bar{a})$ where $a\bar{a} = 1 \mod n$.*

3.5 Solutions to Inline Exercises

Exercise 3.24 (Lemma 3.2)   *Let $\mu: X \to \mathbb{R}^n$ be the moment map of a Hamiltonian T^n-action. If $s: \mathbb{R}^n \to \mathbb{R}$ is a linear map, $s(b_1, \ldots, b_n) = \sum s_i b_i$ then $s \circ \mu$ generates the Hamiltonian flow $\phi_{(s_1 t, \ldots, s_n t)}^{\mu}$.*

Proof   We have $\iota_{V_{\sum_i s_i \mu_i}}\omega = -d(\sum_i s_i \mu_i) = -\sum_i s_i d\mu_i = \sum_i \iota_{s_i V_{\mu_i}}\omega$, and $\sum s_i V_{\mu_i}$ generates the flow $\Phi_{(s_1 t, \ldots, s_n t)}^{\mu}$.   □

Exercise 3.25 (Example 3.11)   *Consider the unit 2-sphere (S^2, ω) where ω is the area form. By comparing infinitesimal area elements, show that the projection map from S^2 to a circumscribed cylinder is area-preserving. Let $H: S^2 \to \mathbb{R}$ be the height function $H(x, y, z) = z$ (thinking of S^2 embedded in the standard way in $\mathbb{R}^3$). Show that H is an action coordinate.*

Solution Let $\hat{n} = (x, y, z)$ be the unit normal vector field to the unit sphere. The area element on the unit sphere is given by $\sigma = \iota_{\hat{n}}(dx \wedge dy \wedge dz) = xdy \wedge dz + ydz \wedge dx + zdx \wedge dy$. If we use cylindrical coordinates $x = r\cos\theta$, $y = r\sin\theta$ then $dx = \cos\theta dr - r\sin\theta d\theta$ and $dy = \sin\theta dr + r\cos\theta d\theta$, so (after some algebra):

$$\sigma = r^2 d\theta \wedge dz + rzdr \wedge d\theta.$$

The unit sphere is defined by the equation $r^2 + z^2 = 1$, which means that $rdr = -zdz$ on the sphere. Therefore

$$\sigma = (1 - z^2)d\theta \wedge dz - z^2 dz \wedge d\theta = d\theta \wedge dz.$$

The unit cylinder has area element $\tau = d\theta \wedge dz$. The projection map from the sphere to the cylinder is $p(r, \theta, z) = (1, \theta, z)$, so $p^*\tau = d\theta \wedge dz = \sigma$.

Observe that the Hamiltonian $H(x, y, z) = z$ gives the Hamiltonian vector field ∂_θ, which rotates the sphere with constant speed so that all orbits have period 2π. Therefore H is an action coordinate (with angle coordinate θ). □

Exercise 3.26 (Lemma 3.23) *The lens space $L(n, a + kn)$ is diffeomorphic to $L(n, a)$ for all integers k. The lens space $L(n, a)$ is diffeomorphic to $L(n, \bar{a})$ where $a\bar{a} = -1 \mod n$.*

Solution Let X be the $\frac{1}{n}(1, a)$ singularity and $\boldsymbol{H} : X \to \mathbb{R}^2$ be the moment map from Example 3.21 with image $\pi(n, a)$. Recall that the lens space $L(n, a)$ is the preimage under $\boldsymbol{H}$ of the horizontal line segment ℓ shown in Figure 3.8.

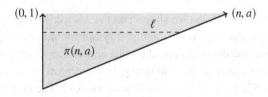

Figure 3.8 The moment polygon $\pi(n, a)$. The preimage of the line ℓ is the lens space $L(n, a)$.

Let X' be the cyclic quotient singularity $\frac{1}{n}(1, a+kn)$, whose moment image is $\pi(n, a+kn)$ shown in Figure 3.9. The integral affine transformation $M = \begin{pmatrix} 1 & k \\ 0 & 1 \end{pmatrix}$ relates these moment polygons: $\pi(n, a)M = \pi(n, a + kn)$.

Since the moment polygons are related by M, Lemma 2.25 gives us a fibred symplectomorphism $X' \to X$. The image of $L(n, a)$ under this fibred symplec-tomorphism lives over the (now slanted) line ℓM. We can isotope ℓM until it is

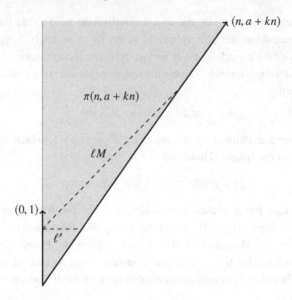

Figure 3.9 The moment polygon $\pi(n, a + kn)$. The lens space $L(n, a + kn)$ living over the line ℓ' is isotopic to the lens space $L(n, a)$ living over the line ℓM.

a horizontal segment ℓ'. The preimages are isotopic, and hence diffeomorphic. The preimage of ℓ' is $L(n, a + kn)$ by definition. Thus $L(n, a) \cong L(n, a + kn)$.

If $a\bar{a} = 1 \mod n$ then $a\bar{a} + tn = 1$ for some t. Let $N = \begin{pmatrix} -a & t \\ n & \bar{a} \end{pmatrix}$. We have $\pi(n, a)N = \pi(n, \bar{a})$, which in turn shows that the associated lens spaces $L(n, a)$ and $L(n, \bar{a})$ are diffeomorphic via the fibred symplectomorphism associated to the integral affine transformation N. If this seems like magic, the trick to finding N is first to reflect $\pi(n, a)$ in the y-axis to get the wedge $\pi(-n, a)$, and then hunt for a matrix in $SL(2, \mathbb{Z})$ which sends $(-n, a)$ to $(0, 1)$. The composite is then $N \in GL(2, \mathbb{Z})$. □

4

Symplectic Reduction

We now introduce *symplectic reduction*, an operation which allows us to construct many interesting symplectic manifolds. A special case of this is *symplectic cut*, which you will use in Exercise 4.42 to construct all toric manifolds.

4.1 Symplectic Reduction

Definition 4.1 Let (X, ω) be a symplectic manifold and let $H: X \to \mathbb{R}$ be a Hamiltonian. Suppose that $\phi_{2\pi}^H(x) = x$ for all $x \in X$. Then the flow defines an action of the circle $S^1 = \mathbb{R}/2\pi\mathbb{Z}$ on X. We this a *Hamiltonian circle action*. We will write $M_c := H^{-1}(c)$ for the level sets of c.

Remark 4.2 Recall that a group action is called *effective* if the only group element which acts as the identity is the identity, and *free* if every point has trivial stabiliser. The quotient of a manifold by a free circle action is again a manifold. If all stabilisers are finite then the quotient is an orbifold.

Here are some of the key facts about Hamiltonian circle actions.

Lemma 4.3 (Exercise 4.35) *Suppose $H: X \to \mathbb{R}$ generates a circle action.*

(a) The critical points of H are precisely the fixed points of the circle action.
(b) The level sets M_c are preserved by the circle action.
(c) If x is a regular point then $T_x M_c = \ker(dH)$.
(d) If $v \in T_x M_c$ satisfies $\omega(v, w) = 0$ for all $w \in T_x M_c$ then $v \in \text{span}(V_H)$.

By Lemma 4.3(a), the stabiliser of a Hamiltonian circle action at a critical point of H is the whole circle. Since any smooth function on a compact manifold has critical points, Hamiltonian circle actions on compact manifolds are never free. For this reason, we restrict attention to a regular level set.

Lemma 4.4   *If c is a regular value of H and $M_c = H^{-1}(c)$ is the regular level set over c then the quotient $Q_c := M_c/S^1$ is an orbifold.*

Proof   By Lemma 4.3(b), the level set is M_c is preserved by the circle action, so the quotient makes sense. We need to show that the stabiliser of the circle action at a point $x \in M_c$ is finite. Since x is regular, it is not a critical point of H, so by Lemma 4.3(a), the stabiliser at x is a proper subgroup of S^1. The circle is compact, and stabilisers are closed subgroups, hence compact. The only proper compact subgroups of S^1 are finite.   □

Lemma 4.5 (Symplectic reduction)   *Suppose $H: X \to \mathbb{R}$ generates a circle action and c is a regular value of H. Suppose for simplicity that the action on M_c is free. Write $i: M_c \to X$ and $p: M_c \to Q_c := M_c/S^1$ for the inclusion and quotient maps respectively. There is a unique symplectic form σ on Q_c such that $i^*\omega = p^*\sigma$. We call (Q_c, σ) the* symplectic quotient *or* symplectic reduction *of X by the Hamiltonian circle action* at level c.

Remark 4.6   One can drop the assumption that the action is free at the cost of allowing quotients which are orbifolds.

Proof   Suppose that $v, w \in T_x M_c$ are tangent vectors to the level set. We want to show that $\omega(v, w)$ depends only on p_*v and $p_*w \in T_{p(x)}Q_c$, so that $\omega(v, w) = \sigma(p_*v, p_*w)$ for some 2-form σ on Q_c. In other words, we want to show that if $x' \in M_c$ is another point with $p(x') = p(x)$ and v', w' are vectors in $T_{x'}M_c$ with $p_*v' = p_*v$ and $p_*w' = p_*w$ then $\omega(v', w') = \omega(v, w)$.

Since $p(x') = p(x)$, we have $\phi_t^H(x) = x'$ for some $t \in S^1$. The vectors v', $(\phi_t^H)_*v$, w', and $(\phi_t^H)_*w$ all live in $T_{x'}M_c$ so we can add/subtract them. Since $p \circ \phi_t^H = p$, we have $p_* \circ (\phi_t^H)_* = p_*$, so

$$p_*(v' - (\phi_t^H)_*v) = p_*v' - p_*v = 0.$$

Similarly $p_*(w' - (\phi_t^H)_*w) = 0$. Since the kernel of p_* is spanned by V_H, we have

$$v' = (\phi_t^H)_*v + aV_H, \qquad w' = (\phi_t^H)_*w + bV_H$$

for some $a, b \in \mathbb{R}$. Thus

$$\omega(v', w') = \omega\left((\phi_t^H)_*v + aV_H, (\phi_t^H)_*w + bV_H\right)$$
$$= \omega(v, w) + a\, dH(v) - b\, dH(w) = \omega(v, w),$$

where we have used the fact that v and w are tangent to a level set, so are annihilated by dH. This shows the existence of σ.

We will now show that σ is nondegenerate. Given a nonzero vector $u \in$

$T_{p(x)}Q_c$, pick a vector $v \in T_x M_c$ with $p_* v = u$. This is possible since p is a submersion. Since the projection $p_* v = u$ is nonzero, v is not a multiple of V_H. By Lemma 4.3(d), there exists $w \in T_x M_c$ such that $\omega(v, w) \neq 0$. Therefore, $\sigma(u, p_* w) \neq 0$, showing that σ is nondegenerate.

The fact that σ is closed follows from Lemma 4.7 below applied to $d\sigma$. □

Lemma 4.7 (Exercise 4.36) *If $p: M \to Q$ is a submersion and η is a k-form on Q such that $p^* \eta = 0$ then $\eta = 0$.*

We finish this section by proving a lemma that will help us to construct Hamiltonian circle or torus actions on symplectic reductions.

Lemma 4.8 *Suppose $G: X \to \mathbb{R}$ Poisson-commutes with H then:*

(a) *$G|_{H^{-1}(c)}$ descends to a function $\overline{G}: H^{-1}(c)/S^1 \to \mathbb{R}$.*
(b) *The Hamiltonian vector field $V_{\overline{G}}$ is equal to[1] $p_* V_G$.*
(c) *If G generates a Hamiltonian circle action on $H^{-1}(c)$ then $\overline{G}$ generates a circle action on the symplectic quotient.*

Proof In this proof, we will write $i: H^{-1}(c) \to X$ for the inclusion of the c-level set.

(a) Since G Poisson-commutes with H, Lemma 1.16 implies it is constant along H-orbits, and hence descends to a function $\overline{G}$ on the quotient, that is, $i^* G = p^* \overline{G}$.

(b) Since G Poisson-commutes with H, the vector field V_G is tangent to $H^{-1}(c)$. This means that the restriction $v := V_G|_{H^{-1}(c)}$ makes sense as a vector field on $H^{-1}(c)$ and $V_G = i_* v$. We want to show that $\iota_{p_* v}\sigma = -d\overline{G}$.

We know that $\iota_{V_G}\omega = -dG$. Pulling back via i gives $\iota_v i^*\omega = -di^* G$, or $\iota_v p^*\sigma = -dp^*\overline{G}$. This implies that $p^*\iota_{p_* v}\sigma = p^*(-d\overline{G})$. Since p is a submersion, Lemma 4.7 implies that $\iota_{p_* v}\sigma = -d\overline{G}$ as required.

(c) Part (b) implies that $\phi_t^{\overline{G}}(p(x)) = p(\phi_t^G(x))$, so if $\phi_{2\pi}^G = $ id then $\phi_{2\pi}^{\overline{G}} = $ id. □

4.2 Examples

Example 4.9 (Complex projective spaces) Let $X = \mathbb{R}^{2n}$ with coordinates $(x_1, y_1, \ldots, x_n, y_n)$ and symplectic form $\sum_i dx_i \wedge dy_i$. Consider the Hamiltonian $H = \frac{1}{2}\sum_i(x_i^2 + y_i^2)$. The Hamiltonian vector field is $(-y_1, x_1, \ldots, -y_n, x_n)$, which generates a circle action rotating each xy-plane at constant angular speed. The nonempty regular level sets are the spheres M_c of radius $\sqrt{2c}$. If $c > 0$ we

[1] Here, we are abusively identifying V_G with its restriction to $H^{-1}(c)$.

get a nonempty regular level set, whose symplectic quotient is called *complex projective space* $\mathbb{CP}^{n-1}$. When $c = 1$, we call the reduced symplectic form on $\mathbb{CP}^{n-1}$ the *Fubini–Study form* ω_{FS}. This is normalised so that the moment image is a simplex with edges of affine length 1, which means that the complex lines in $\mathbb{CP}^{n-1}$ have symplectic area 2π. (Exercise 4.37: Check directly from this definition of ω_{FS} on $\mathbb{CP}^1$ that $\int_{\mathbb{CP}^1} \omega_{FS} = 2\pi$.)

Remark 4.10   If $c < 0$ then the symplectic quotient is empty. In general, it is an interesting problem to understand how the topology of the symplectic quotient varies when the parameter c crosses a critical value; this is the subject of *symplectic birational geometry* [50, 67].

Remark 4.11   We can identify $\mathbb{R}^{2n}$ with $\mathbb{C}^n$ by introducing complex coordinates $z_k = x_k + iy_k$. Then the circle action generated by H is precisely the action $\mathbf{z} \mapsto e^{it}\mathbf{z}$. The orbits (other than the origin) are precisely the circles of fixed radius in the complex lines of $\mathbb{C}^n$, so our symplectic quotient coincides with the usual definition[2] of $\mathbb{CP}^{n-1}$ as the space of complex lines through the origin in $\mathbb{C}^n$. The advantage of defining it as a symplectic quotient is the clean construction of ω_{FS}; we did not need to write down an explicit Kähler potential.

In fact, we also recover the standard torus action on $\mathbb{CP}^{n-1}$ using Lemma 4.8. Consider the Hamiltonian system $G = (G_1, \ldots, G_n) \colon \mathbb{C}^n \to \mathbb{R}^n$ where $G_k = \frac{1}{2}|z_k|^2$. These all commute with the Hamiltonian $H = \sum_{k=1}^n G_k$ from Example 4.9. Therefore they descend to give a Hamiltonian system

$$(\overline{G}_1, \ldots, \overline{G}_n) \colon \mathbb{CP}^{n-1} \to \mathbb{R}^n.$$

The image of this Hamiltonian system is simply the image of $G|_{H^{-1}(c)}$, which is the intersection of the hyperplane $\sum G_k = c$ with the nonnegative orthant in $\mathbb{R}^n$, that is, an $(n-1)$-simplex (Figure 4.1).

Example 4.12 (Weighted projective spaces)   Let $a_1, \ldots, a_n$ be positive integers and consider the Hamiltonian $H \colon \mathbb{C}^n \to \mathbb{R}$ given by

$$H(z_1, \ldots, z_n) = \frac{1}{2} \sum_{k=1}^n a_k |z_k|^2.$$

This generates the circle action $z_k \mapsto e^{ia_k t} z_k$. The symplectic quotient is an orbifold called the *weighted projective space* $\mathbb{P}(a_1, \ldots, a_n)$. If $a_k \neq 1$ then the points of the form $(0, \ldots, 0, z_k, 0, \ldots, 0)$ have nontrivial stabiliser $\{\mu \in S^1 : \mu^{a_k} = 1\}$. If $\gcd(a_k, a_\ell) = 1$ then points with z_k and z_ℓ both nonzero have trivial stabiliser, so if we assume that the a_k are pairwise coprime then the only

---

[2]   If you are unfamiliar with this description of complex projective space and with the role it plays in algebraic geometry, you can read more in Appendix C.

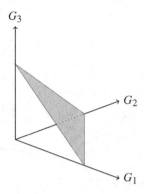

Figure 4.1 The moment image of $\mathbb{CP}^2$ in $\mathbb{R}^3$ when considered as a symplectic quotient of $\mathbb{C}^3$.

singularities are the isolated cyclic quotient singularities with all but one of the z_k equal to zero. As in the case of $\mathbb{CP}^{n-1}$, the functions $\frac{1}{2}|z_k|^2$ descend and generate a torus action on $\mathbb{P}(a_1, \ldots, a_n)$, whose moment polytope is the simplex in $\mathbb{R}^n$ with vertices at

$$(c/a_1, 0, \ldots, 0), \ (0, c/a_2, 0, \ldots, 0), \ \ldots, \ (0, \ldots, 0, c/a_n).$$

In Figure 4.2(a), you can see the moment triangle for $\mathbb{P}(1, 2, 3)$. If you project it onto the yz-plane, you get the triangle shown in Figure 4.2(b). This has a smooth point over the Delzant vertex and two singularities modelled on the $\frac{1}{2}(1, 1)$ and $\frac{1}{3}(1, 2)$ singularities.

Example 4.13 Consider the round metric on the n-dimensional unit sphere S^n; in particular, geodesics are great circles, and if you traverse a great circle with constant speed 1 then it returns to its starting point after time 2π. Let $H: T^*S^n \to \mathbb{R}$ be the Hamiltonian $\frac{1}{2}|\eta|^2$ generating the cogeodesic flow.[3] The cogeodesic flow does not define a circle action on T^*S^n, but the periodicity of geodesics with speed 1 means that the flow does define a circle action on the level set $H^{-1}(1/2)$ (on which geodesics move with speed 1). This allows us to perform symplectic reduction at that level. The result is that the space of (oriented) geodesics parametrised by arc-length on the round S^n is naturally a symplectic manifold of dimension $2n - 2$. We can also identify this manifold of oriented geodesics as a homogeneous space: the great circles are intersections of $S^n \subseteq \mathbb{R}^{n+1}$ with oriented 2-planes through the origin, so the manifold of

---

[3] If you are unfamiliar with cotangent bundles and cogeodesic flow, you can read more about them in Appendix D.

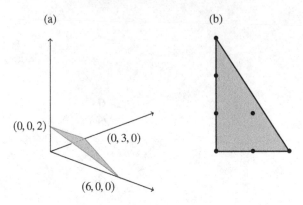

Figure 4.2 (a) The moment image of the weighted projective space $\mathbb{P}(1, 2, 3)$ in $\mathbb{R}^3$ when considered as a symplectic quotient of $\mathbb{C}^3$. The coordinates of the vertices of the triangle are given assuming the symplectic reduction is at level $c = 6$.
(b) The projection of this triangle to the yz-plane, with integer lattice points marked.

oriented geodesics coincides with the Grassmannian $\widetilde{Gr}(2, n + 1)$ of oriented 2-planes in $\mathbb{R}^{n+1}$.

Remark 4.14 (Exercise 4.38)   One can identify the Grassmannian $\widetilde{Gr}(2, n + 1)$ with the homogeneous space $O(n + 1)/(SO(2) \times O(n - 1))$, or with the quadric hypersurface $\sum_{i=1}^{n+1} z_i^2 = 0$ in $\mathbb{CP}^n$ with homogeneous coordinates[4] $[z_1 : \cdots : z_{n+1}]$.

Remark 4.15 (Exercise 4.39)   The zero-section $S^n \subseteq T^*S^n$ consists of fixed points of the cogeodesic flow, but all the other orbits are circles. By a similar argument to Theorem 1.4, one can modify $H = \frac{1}{2}|\eta|^2$ away from the zero-section to get a Hamiltonian for which all orbits have the same period (i.e. giving a circle action away from the zero-section). What Hamiltonian should you take instead?

4.3 Symplectic Cut

The symplectic cut is a particularly useful case of symplectic reduction, introduced by Lerman [65]. Here is a simple example to illustrate the operation.

Example 4.16   Consider the cylinder $X = \mathbb{R} \times S^1$ with symplectic form $dp \wedge dq$

---

[4] Homogeneous coordinates are introduced in Appendix C.

(where q is the angular coordinate on S^1). The function $H = p$ generates the circle action which rotates the S^1 factor, and the symplectic reduction at any level c yields a single point. If we symplectically reduce at level c but leave all the other levels alone then this has the effect of 'pinching' the cylinder along a circle. Although the result is singular, it is the union of two smooth symplectic discs, which we will denote by $\overline{X}_{H \leq c}$ and $\overline{X}_{H \geq c}$. Each of these contains the point $H^{-1}(c)/S^1$ at its centre. We will adopt the convention of calling $\overline{X}_{H \geq c}$ *the symplectic cut of X at level c.*

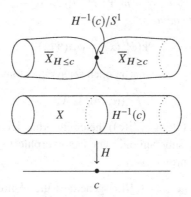

Figure 4.3 Symplectic cut. It is as if we have cut X along $H^{-1}(c)$ and sewn up the wound by collapsing the circle-orbit to a point.

We can perform symplectic cuts in a completely systematic and general way. Let (X, ω) be a symplectic manifold and suppose that $H \colon X \to \mathbb{R}$ is a Hamiltonian generating a circle action ϕ_t^H. Pick a regular value c for which the circle action on $H^{-1}(c)$ is free. Consider the product $X \times \mathbb{C}$ with the symplectic form $\omega + dp \wedge dq$ (where $p + iq \in \mathbb{C}$) and define the Hamiltonian $\tilde{H}_c(x, \xi) = H(x) - c - \frac{1}{2}|\xi|^2$.

Lemma 4.17 (Exercise 4.40)   *The Hamiltonian $\tilde{H}_c$ generates the circle action*

$$(x, \xi) \mapsto \left(\phi_t^H(x), e^{-it}\xi \right)$$

on $X \times \mathbb{C}$.

Definition 4.18 (Symplectic cut at level c)   The symplectic cut $\overline{X}_{H \geq c}$ of X at level c is defined to be the symplectic reduction of $X \times \mathbb{R}^2$ with respect to the Hamiltonian $\tilde{H}_c$ at level zero. One can also define a symplectic cut $\overline{X}_{H \leq c}$ by using the Hamiltonian $H - c + \frac{1}{2}|\xi|^2$. We will write $[x, \xi]$ for the equivalence class of $(x, \xi) \in \tilde{H}_c^{-1}(0)$ in the symplectic cut.

We now try to understand what $\overline{X}_{H \geq c}$ looks like.

Lemma 4.19 *Consider the function* $X \times \mathbb{C} \to \mathbb{R}$, $(x, \xi) \mapsto H(x)$. *This Poisson-commutes with* $\tilde{H}_c$ *and hence (by Lemma 4.8) descends to a function* $\overline{H} \colon \overline{X}_{H \geq c} \to \mathbb{R}$ *which generates a circle action on* $\overline{X}_{H \geq c}$. *Moreover:*

(a) *The image of* $\overline{H}$ *is contained in* $[c, \infty)$.

(b) *The preimage* $\overline{H}^{-1}(c)$ *is symplectomorphic to the symplectic reduction* $H^{-1}(c)/S^1$.

(c) *There is a symplectomorphism* $\Psi \colon H^{-1}((c, \infty)) \to \overline{H}^{-1}((c, \infty))$ *which intertwines the circle actions:*

$$\Psi(\phi_t^H(x)) = \phi_t^{\overline{H}}(\Psi(x)).$$

In other words, $\overline{X}_{H \geq c}$ contains an open set symplectomorphic to

$$H^{-1}((c, \infty)) \subseteq X.$$

Where it has been cut (along $H^{-1}(c)$) the circle-orbits are collapsed to points, yielding a symplectic submanifold symplectomorphic to $H^{-1}(c)/S^1$. Everything below level c is thrown away.

Proof The function $(x, \xi) \mapsto H(x)$ generates the Hamiltonian vector field $(V_H, 0)$ on $X \times \mathbb{C}$. Recall that $\tilde{H}_c = H - c + \frac{1}{2}|\xi|^2$. Since $dH(V_H, 0) = 0$ and $d|\xi|^2(V_H, 0) = 0$ we have $\{H, \tilde{H}_c\} = -d\tilde{H}_c(V_H, 0) = 0$. By Lemma 4.8(a), this function descends to give a function $\overline{H}$ on the symplectic quotient. Since $(V_H, 0)$ generates a circle action on $X \times \mathbb{C}$, Lemma 4.8(c) implies that $\overline{H}$ generates a circle action on $\overline{X}_{H \geq c}$.

Property (a): If $(x, \xi) \in \tilde{H}_c^{-1}(0)$ then $H(x) - c + \frac{1}{2}|\xi|^2 = 0$, so

$$H(x) = c + \frac{1}{2}|\xi|^2 \geq c.$$

Therefore $\overline{H}([x, \xi]) = H(x) \geq c$.

Property (b): The map $H^{-1}(c)/S^1 \to \overline{H}^{-1}(c) \subseteq \overline{X}_{H \geq c}$ defined by $[x] \mapsto [x, 0]$ is the required symplectomorphism.

Property (c): The required symplectomorphism is

$$\Psi(x) = \left[x, \sqrt{2(H(x) - c)} \right]. \qquad \square$$

4.4 Further Examples

Most of our examples will be based on the following special case.

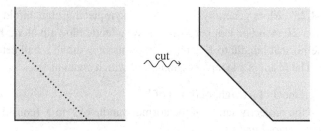

Figure 4.4 The moment polygon for symplectic blow-up of $\mathbb{C}^2$ at the origin.

Example 4.20 Suppose that X is a toric manifold and that $\mu\colon X \to \mathbb{R}^n$ is the moment map with moment image $\Delta = \mu(X)$. Pick a $\mathbb{Z}$-linear map $s\colon \mathbb{R}^n \to \mathbb{R}$ and take $H = s \circ \mu$. If we symplectically cut at level c then the result is still toric by Lemma 4.8, and the moment image is given by

$$\{x \in \Delta \,:\, s(x) \geq c\}.$$

In other words, this is the truncation of the moment image by the half-space $\{s \geq c\}$.

Remark 4.21 (Exercise 4.41)   We need s to be $\mathbb{Z}$-linear (or at least $\mathbb{Q}$-linear) for this construction to work. Can you explain why?

Theorem 4.22 (Exercise 4.42)   *Any convex rational polytope Δ occurs as the moment image of a toric Hamiltonian system (on a possibly singular space).*

Example 4.23 Take $X = \mathbb{C}^n$ and $H(z) = \frac{1}{2}\sum_{k=1}^{n} |z_k|^2$. We saw in Example 4.9 that the symplectic *reduction* at level $c > 0$ is $\mathbb{CP}^{n-1}$. If instead we take the symplectic *cut* at this level, we get a toric manifold whose moment image is

$$\left\{b \in \mathbb{R}^n_{\geq 0} \,:\, \sum b_k \geq c\right\}.$$

(See Figure 4.4 for the $n = 2$ case.) This contains the complement of the symplectic ball $B_c := \{z \in \mathbb{C}^n \,:\, |z|^2 \leq 2c\}$ and it also contains a copy of $\mathbb{CP}^{n-1}$ living over the newly cut facet $\sum b_k = c$. This is known as the *symplectic blow-up* of X in the ball B_c. More generally, you can perform this operation on any Delzant vertex of a moment polytope (see Example 4.26). The symplectic area of the exceptional curve is $2\pi c$.

Remark 4.24   Compare Figure 4.4 with Figure 3.4. This shows that symplectically blowing up a ball is essentially the same as the complex blow-up of a smooth point, though, of course, they happen in different categories, so one must be careful when relating them in practice. A paper which very carefully explains (and uses) the relationship is [75, Section 2].

Example 4.25   More generally, if $Y \subseteq X$ is a symplectic submanifold of real codimension $2k$ then one can perform this symplectic blow-up along Y: each fibre of the normal bundle to Y is replaced by its blow-up in a ball centred at the origin. The result is a symplectic manifold which contains:

- the complement of a neighbourhood of Y
- a copy of the projectivisation of the normal bundle of Y in X (considered as a complex vector bundle).

For more details, see Usher's MathOverflow answer [110].

Example 4.26 (Exercise 4.43)   There is a blow-up of $\mathbb{CP}^1 \times \mathbb{CP}^1$ in two disjoint symplectic balls which is symplectomorphic to a blow-up of $\mathbb{CP}^2$ in three disjoint balls.

Example 4.27 (Exercise 4.44)   Show that the common blow-up from Example 4.26 arises as a symplectic reduction of $\mathbb{CP}^1 \times \mathbb{CP}^1 \times \mathbb{CP}^1$.

Example 4.28   As in Example 4.23, we take $X = \mathbb{C}^n$ and $H(z) = \frac{1}{2} \sum_{k=1}^n |z_k|^2$, but this time we use the symplectic cut $\overline{X}_{H \leq c}$. The result is a symplectic toric manifold whose moment image is the n-simplex $\{ \boldsymbol{b} \in \mathbb{R}^n \ : \ b_k \geq 0, \sum b_k \leq c \}$. When $c = 1$, we saw this moment polytope arise in Example 3.17 as the moment image of the standard torus action on $\mathbb{CP}^n$, so $\overline{X}_{H \leq 1}$ is symplectomorphic to $\mathbb{CP}^n$ with the Fubini–Study form. More generally, if $c > 0$, the only difference is that the symplectic form is rescaled by a factor of c.

The previous example illustrates how the standard operation of *projective compactification* in algebraic geometry can be understood using symplectic cut. We can do something similar starting with Example 4.13.

Example 4.29   Consider the Hamiltonian $H \colon T^* S^n \to \mathbb{R}$ from Example 4.13. The symplectic cut $\overline{(T^* S^n)}_{H \leq 1/2}$ is a compact symplectic manifold containing an open subset symplectomorphic to the open unit cotangent bundle of S^n and a 'compactifying divisor' $H^{-1}(1/2)/S^1$ symplectomorphic to the Grassmannian $\widetilde{Gr}(2, n+1)$. In fact, $\overline{(T^* S^n)}_{H \leq 1/2}$ is symplectomorphic to a quadric hypersurface in $\mathbb{CP}^{n+1}$ and the compactifying divisor is a hyperplane section. For an in-depth discussion of this and similar examples, see [6].

4.5 Resolution of Singularities

We have seen that symplectic cut can be used as a symplectic Ersatz for blowing up in algebraic geometry. Just as blowing up allows us to resolve singularities in algebraic geometry, symplectic cut can be used to remove singularities in symplectic geometry.

Example 4.30 Consider the cyclic quotient singularity $\frac{1}{a_0}(1, a_1)$ with moment polygon $\pi(a_0, a_1)$ from Example 3.21. We can symplectically cut using a horizontal slice at a level above the singularity to obtain a polygon $\tilde{\pi}(a_0, a_1)$ which now has two vertices: a Delzant corner and a corner modelled on $\pi(a_1, a_2)$ where $0 \leq a_2 < a_1$ satisfies $a_2 = -a_0 \mod a_1$, that is, $a_2 = y_1 a_1 - a_0$ for some y_1. Namely, the matrix $M_1 = \begin{pmatrix} 0 & -1 \\ 1 & y_1 \end{pmatrix}$ satisfies $(-1, 0)M_1 = (0, 1)$, $(a_0, a_1)M_1 = (a_1, a_2)$. See Figure 4.5.

If $a_2 = 1$ then this second corner is Delzant, and the symplectic cut is smooth. Otherwise, we can iterate this procedure by slicing the polygon $\pi(a_1, a_2)$ horizontally (i.e. slicing $\tilde{\pi}(a_0, a_1)$ parallel to $(1, 0)M_1^{-1}$). We get a decreasing sequence of positive integers $a_0, a_1, a_2, \ldots$ and a sequence of positive integers $y_1, y_2, \ldots$ with $a_{k+1} = y_k a_k - a_{k-1}$. At some point we necessarily find $a_{m+1} = 0$ because the sequence cannot continue decreasing forever. The result of all these cuts is a Delzant polygon. See Figures 4.5 and 4.6.

Unpacking the recursion formula for a_{k+1}, we see that

$$a_0/a_1 = y_1 - a_2/a_1 = y_1 - \frac{1}{a_1/a_2} = y_1 - \frac{1}{y_2 - a_3/a_2} = \cdots = y_1 - \cfrac{1}{y_2 - \cfrac{1}{\ddots - \cfrac{1}{y_m}}}.$$

Thus, the numbers y_i are the coefficients in the continued fraction expansion[5] of a_0/a_1.

This process has introduced m compact edges into our polygon. There are symplectic spheres living over these in the iterated symplectic cut. These spheres have self-intersection numbers $-y_1, -y_2, \ldots, -y_m$. To see this, it suffices to check it for y_1 because of the iterative nature of the process. There are two cases to consider:

- If our process terminated after a single cut then $a_0/a_1 = y_1$ is an integer, so $a_1 = 1$ and $a_0 = y_1$. In this case, the compact edge of $\tilde{\pi}(a_0, a_1)$ has rays emanating in the $(0, 1)$ and $(a_0, a_1) = (y_1, 1)$ directions from its endpoints. By Lemma 3.20, the corresponding sphere has self-intersection $-y_1$.

- If our process takes two or more steps then the leftmost compact edge has outgoing rays pointing in the $(0, 1)$ and $(1, 0)M_1^{-1} = (y_1, 1)$ directions. Again, by Lemma 3.20, the corresponding sphere has self-intersection $-y_1$.

---

[5] This is the *Hirzebruch–Jung* convention for taking continued fractions with minus signs.

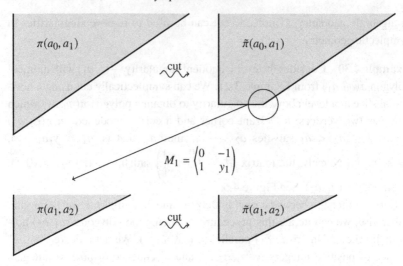

Figure 4.5 The minimal resolution of cyclic quotient surface singularities via symplectic cuts. In this specific example we have used $a_0 = 3$, $a_1 = 2$, $a_2 = 1$, $y_1 = y_2 = 2$ (so the process terminates after two cuts: the result is shown in Figure 4.6).

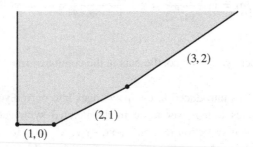

Figure 4.6 The combined result of performing both cuts in Figure 4.5; vertices are marked with dots for emphasis. In this example (the $\frac{1}{3}(1,2)$ singularity) the vectors indicate the directions of the edges and, using Lemma 3.20, you can check that we have introduced two -2-spheres in the minimal resolution.

This process allows us to replace a $\frac{1}{a_0}(1, a_1)$ singularity by a chain of symplectic spheres with self-intersections determined by the continued fraction expansion of a_0/a_1. This is precisely what happens when we perform the *minimal resolution* of this singularity in complex geometry, so we often refer to this sequence of symplectic cuts as taking the minimal resolution. Figures 4.5

and 4.6 illustrate this process in the case $a_0 = 3$, $a_1 = 2$. Since $\frac{3}{2} = 2 - \frac{1}{2}$, the minimal resolution replaces the $\frac{1}{3}(1,2)$ singularity by two -2-spheres.

Remark 4.31   Just as symplectic blow-up involves a choice of parameter (the area of the exceptional sphere), forming the minimal resolution by symplectic cuts involves choices of edge-lengths for the symplectic cuts. This subtlety is absent from the minimal resolution in complex geometry, but appears when you try to equip the minimal resolution with an ample line bundle or Kähler form.

In higher dimensions, resolution of singularities is more complicated. We consider just one six-dimensional example to give a flavour of what can happen.

Example 4.32 (The conifold)   Consider the affine variety $C := \{z_1 z_4 = z_2 z_3\} \subseteq \mathbb{C}^4$. This has a singularity at the origin which goes by many names (A_1 *singularity, node, ordinary double point, conifold*). It admits a Hamiltonian T^3-action:

$$\left(e^{it_1} z_1, e^{it_2} z_2, e^{it_3} z_3, e^{i(t_2+t_3-t_1)} z_4 \right).$$

The moment map for this action is

$$\mu(z_1, \ldots, z_4) = \left(\frac{1}{2}(|z_1|^2 - |z_4|^2), \frac{1}{2}(|z_2|^2 + |z_4|^2), \frac{1}{2}(|z_3|^2 + |z_4|^2) \right).$$

If we write H_1, H_2, H_3 for these three functions then we have

$$H_2 \geq 0, \quad H_3 \geq 0, \quad H_1 + H_2 \geq 0, \quad H_1 + H_3 \geq 0.$$

These four inequalities cut out a polyhedral cone Δ in $\mathbb{R}^3$, spanned by the four rays

$$(r, 0, 0), \quad (0, r, 0), \quad (0, 0, r), \quad (-r, r, r), \quad r > 0.$$

(See Figure 4.7). These rays are the moment images of the curves

$$(z, 0, 0, 0), \quad (0, z, 0, 0), \quad (0, 0, z, 0), \quad (0, 0, 0, z)$$

all of which are contained in C, so $\mu(C)$ contains all four of these rays and hence their convex hull, which is the whole of Δ. Thus, $\mu(C) = \Delta$. Note that Δ is not Delzant at the origin (corresponding to the nodal singularity of C).

One can find some resolutions of this nodal 3-fold by taking symplectic cuts.

Example 4.33 (The fully resolved conifold)   Take the symplectic cut of the conifold C with respect to the Hamiltonian $\frac{1}{2} \sum_{k=1}^{4} |z_k|^2$ at level $c = 1$. In terms of our functions H_1, H_2, H_3 this is just $H_1 + H_2 + H_3$, so we obtain the polytope shown in Figure 4.8. We have introduced a new quadrilateral facet $H_1 + H_2 + H_3 = 1$. This quadrilateral has vertices at $(1, 0, 0)$, $(0, 1, 0)$,

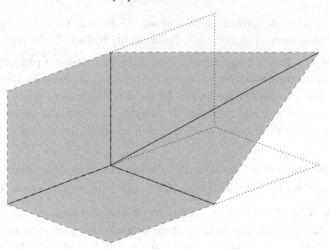

Figure 4.7 The moment polytope for the nodal 3-fold in Example 4.32 is an infinite cone extending this figure.

$(0, 0, 1)$, and $(-1, 1, 1)$. In fact, this is isomorphic to a square under the integral affine transformation $(x, y, z) \mapsto (y, z)$, so the preimage of our new facet is the toric manifold associated to a square. In Example 3.13, we saw that this is symplectomorphic to $\mathbb{CP}^1 \times \mathbb{CP}^1$. We have replaced the nodal point of C with an 'exceptional divisor' $\mathbb{CP}^1 \times \mathbb{CP}^1$. Note that this is precisely what we would get if we treated $z_1, \ldots, z_4$ as homogeneous coordinates: $z_1 z_4 = z_2 z_3$ is a smooth projective quadric surface, and hence biholomorphic to $\mathbb{CP}^1 \times \mathbb{CP}^1$ (see Example F.3). In the language of algebraic geometry, we have performed a blow-up of $\mathbb{C}^4$ at the origin and taken the proper transform of C.

Example 4.34 (Small-resolved conifold)   Take a linear map s which vanishes along one of the facets of Δ, for example $s(x, y, z) = z$. Take the symplectic cut $\overline{C}_{s \circ \mu \geq \epsilon}$ of the conifold for some small $\epsilon > 0$. If Δ were Delzant, this truncation would have no effect on the combinatorics of the moment polytope and would just change the lengths of some edges. Because Δ is not Delzant, the result is to introduce a new edge (see Figure 4.9). This is an example of a *small resolution*: the symplectic cut is a smooth manifold, but instead of replacing the singularity with a divisor, we have replaced it with a curve. In algebro-geometric language, we have blown up the *Weil divisor* $\{z_3 = z_4 = 0\} \subseteq C$. Roughly speaking, a Weil divisor is a complex codimension 1 submanifold; if a Weil divisor can be defined by the vanishing of a single polynomial then it is called a *Cartier*

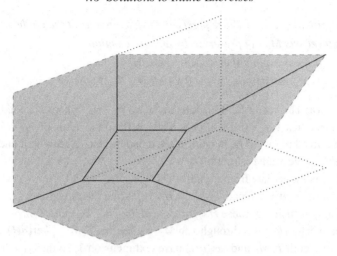

Figure 4.8 The moment polytope of the fully resolved conifold in Example 4.33. This is obtained by intersecting the polytope from Figure 4.7 with the half-space $x + y + z \geq 1$.

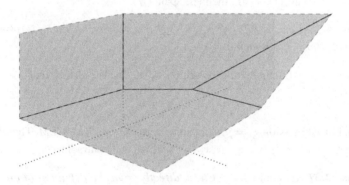

Figure 4.9 The moment polytope of the small-resolved conifold in Example 4.32. This is obtained by intersecting the polytope in Figure 4.7 with the half-space $z \geq \epsilon$.

divisor. Blowing up along a Cartier divisor has no effect, but we have blown up along a non-Cartier divisor (requiring both z_3 and z_4 to vanish).

4.6 Solutions to Inline Exercises

Exercise 4.35 (Lemma 4.3) *Suppose $H: X \to \mathbb{R}$ generates a circle action.*

(a) The critical points of H are precisely the fixed points of the circle action.

(b) The level sets M_c are preserved by the circle action.

(c) If x is a regular point then $T_x M_c = \ker(dH)$.

(d) If $v \in T_x M_c$ satisfies $\omega(v, w) = 0$ for all $w \in T_x M_c$ then $v \in \mathrm{span}(V_H)$.

Solution   (a) The vector field V_H satisfies $\iota_{V_H} \omega = -dH$. Since ω is nondegenerate, we see that $V_H = 0$ if and only if $dH = 0$, so the zeros of V_H coincide with the critical points of H. A point is fixed under the circle action if and only if the vector field vanishes there.

(b) This is immediate from Lemma 1.11.

(c) Suppose that γ is a path in $M_c = H^{-1}(c)$. Then the directional derivative of H along $\dot{\gamma}$ vanishes because $H \circ \gamma = c$, so $dH(\dot{\gamma}) = 0$. Because $T_x M_c$ consists of tangent vectors to paths through x in M_c, we see that $T_x M_c \subseteq \ker(dH)$. Since x is regular, both $T_x M_c$ and $\ker(dH)$ have codimension 1, so the containment $T_x M_c \subseteq \ker(dH)$ implies they coincide.

(d) The symplectic orthogonal complement $(T_x M_c)^\omega$ is one-dimensional and contains the span of V_H, therefore it equals the span of V_H. This means that if $\omega(v, w) = 0$ for all $w \in T_x M_c$ then $v \in \mathrm{span}(V_H)$.   □

Exercise 4.36 (Lemma 4.7)   *If $p \colon M \to Q$ is a submersion and η is a k-form on Q such that $p^* \eta = 0$ then $\eta = 0$.*

Solution   If $\eta \neq 0$ then there exist vectors $\xi_1, \ldots, \xi_k \in TQ$ such that

$$\eta(\xi_1, \ldots, \xi_k) \neq 0.$$

Since p is a submersion, $\xi_i = p_* v_i$ for some vectors $v_1, \ldots, v_k \in TM$. Therefore, $0 \neq \eta(\xi_1, \ldots, \xi_k) = (p^* \eta)(v_1, \ldots, v_k) = 0$, which is a contradiction.   □

Exercise 4.37 (Example 4.9)   *Check directly from the definition of ω_{FS} on $\mathbb{CP}^1$ that $\int_{\mathbb{CP}^1} \omega_{FS} = 2\pi$.*

Proof   Recall that ω_{FS} comes from symplectically cutting $\mathbb{C}^2$ at radius $r = \sqrt{2}$. Consider the affine coordinate patch $\{[x + iy : 1] : x + iy \in \mathbb{C}\}$ on $\mathbb{CP}^1$. This covers all but a point of $\mathbb{CP}^1$, and we can pick a section of the symplectic quotient $p \colon S^3(r) \to \mathbb{CP}^1$ living over this patch, for example $[x + iy : 1] \mapsto (\frac{r(x+iy)}{1+x^2+y^2}, \frac{r}{1+x^2+y^2}) \in S^3(r) \subseteq \mathbb{C}^2$. The integral of ω_{FS} over this coordinate patch is given by the integral of $\omega_{\mathbb{C}^2}$ over the section (since $p^* \omega_{FS} = i^* \omega_{\mathbb{C}^2}$ where i is the inclusion $S^3(r) \to \mathbb{C}^2$). The form $\omega_{\mathbb{C}^2}$ is the sum $pr_1^* \omega_{\mathbb{C}} + pr_2^* \omega_{\mathbb{C}}$ where pr_1 and pr_2 are the projections to the first and second factors. The projection of our section to the second factor is one-dimensional, so $pr_2^* \omega_{\mathbb{C}}$ integrates trivially over the section. The image of the section under pr_1 is the disc of radius r, so the integral of $pr_1^*\omega_{\mathbb{C}}$ over the section is πr^2. Since $r = \sqrt{2}$, this gives area 2π, as required.   □

Exercise 4.38 (Remark 4.14)   *One can identify the Grassmannian $\widetilde{Gr}(2, n+1)$ with the homogeneous space $O(n + 1)/(SO(2) \times O(n - 1))$, or with the quadric hypersurface $\sum_{i=1}^{n+1} z_i^2 = 0$ in $\mathbb{CP}^n$ with homogeneous coordinates $[z_1 : \cdots : z_{n+1}]$.*

Solution   There is a transitive action of $O(n + 1)$ on oriented 2-planes in $\mathbb{R}^{n+1}$. By the orbit-stabiliser theorem,[6] this means that $\widetilde{Gr}(2, n + 1) = O(n + 1)/\mathrm{Stab}(\mathbb{R}^2)$, where $\mathrm{Stab}(\mathbb{R}^2)$ is the subgroup of $O(n+1)$ stabilising the standard oriented 2-plane $\{(x_1, x_2, 0, \ldots, 0) : x_1, x_2 \in \mathbb{R}\}$. This stabiliser consists of block-matrices $\begin{pmatrix} A & 0 \\ 0 & B \end{pmatrix}$ with $A \in SO(2)$ and $B \in O(n - 1)$.

To identify this homogeneous space with the quadric hypersurface, observe that the quadric also admits an action of $O(n+1)$ (inherited from $\mathbb{C}^{n+1}$) precisely because the quadratic form $\sum_{k=1}^{n+1} z_k^2$ is preserved by orthogonal matrices. This is (a) transitive and (b) the stabiliser of the point $[1 : i : \cdots : 0]$ is isomorphic to $SO(2) \times O(n - 1)$.

(a) To see transitivity, let us prove that the orbit of $[1 : i : 0 : \cdots : 0]$ is the whole quadric. Suppose that $z = x + iy \in \mathbb{C}^{n+1}$ is a nonzero complex vector with $\sum_{k=1}^{n+1} z_k^2 = 0$. The real and imaginary parts of this condition become $|x|^2 = |y|^2$ and $x \cdot y = 0$. Use the Gram–Schmidt process to extend $\hat{x} = x/|x|$, $\hat{y} = y/|y|$ to an orthonormal basis of $\mathbb{R}^{n+1}$ and use these basis vectors as the columns of an orthogonal matrix A. By construction, $A(1, i, 0, \ldots, 0) = \hat{x} + i\hat{y}$, so $A[1 : i : 0 : \cdots : 0] = [\hat{z}] = [z]$. This shows that $[z]$ lies in the $O(n+1)$-orbit of $[1 : i : 0 \cdots : 0]$.

(b) To understand the stabiliser, suppose that $A[1 : i : 0 : \cdots : 0] = [1 : i : 0 : \cdots : 0]$. If the columns of A are $a_1, \ldots, a_{n+1}$ then this condition becomes $a_1 + ia_2 = re^{i\theta}(1, i, 0, \ldots, 0)$ for some $re^{i\theta} \in \mathbb{C} \setminus \{0\}$. Since A is orthogonal, we get $r = 1$ and

$$a_1 = \begin{pmatrix} \cos\theta \\ -\sin\theta \\ 0 \\ \vdots \\ 0 \end{pmatrix}, \qquad a_2 = \begin{pmatrix} \sin\theta \\ \cos\theta \\ 0 \\ \vdots \\ 0 \end{pmatrix}.$$

The upper-right 2-by-2 block of A is therefore an element of $SO(2)$. Orthogonal-

---

[6]   That is, the theorem which identifies the G-orbit of x with $G/\mathrm{Stab}(x)$, in whatever category you are working, e.g. differentiable manifolds and smooth actions.

ity of A now implies that A is block-diagonal and the lower-left $(n-1)$-by-$(n-1)$ block is orthogonal. Thus, $A \in SO(2) \times O(n-1)$.   □

Exercise 4.39 (Remark 4.15) *The zero-section $S^n \subseteq T^*S^n$ consists of fixed points of the cogeodesic flow, but all the other orbits of the cogeodesic flow on the round sphere are circles. By a similar argument to Theorem 1.4, one can modify $H = \frac{1}{2}|\eta|^2$ away from the zero-section to get a Hamiltonian for which all orbits have the same period (i.e. giving a circle action away from the zero-section). What Hamiltonian should you take instead?*

Solution The geodesics in the level set $H^{-1}(c)$ have speed $|p| = \sqrt{2c}$, so have period $T(c) = 2\pi/\sqrt{2c}$. The proof of Theorem 1.4 tells us to use the Hamiltonian $\alpha \circ H$ where $\alpha(b) = \frac{1}{2\pi} \int_0^b T(c)\,dc = \frac{1}{\sqrt{2}} \int_0^b \frac{dc}{\sqrt{c}} = \sqrt{2b}$. In other words, use the Hamiltonian $|p|$. This Hamiltonian is not smooth at $p = 0$ (i.e. along the zero-section) so the new Hamiltonian flow only makes sense away from the zero-section.   □

Exercise 4.40 (Lemma 4.17) *The Hamiltonian $\tilde{H}_c$ generates the circle action*

$$(x, \xi) \mapsto \left(\phi_t^H(x), e^{-it}\xi \right)$$

on $X \times \mathbb{C}$.

Solution The Hamiltonian vector field is $V_H - \partial_\theta$ where θ is the angular coordinate on $\mathbb{C}$. The flowlines are therefore $(\phi_t^H(x), e^{-it}\xi)$, which all have period 2π.   □

Exercise 4.41 (Remark 4.21) *Why do we need s to be $\mathbb{Z}$-linear (or at least $\mathbb{Q}$-linear) for the construction in Example 4.20 to work?*

Solution If s is $\mathbb{Z}$-linear then the Hamiltonian $H := s \circ \mu$ generates a circle action: the Hamiltonian vector field V_H is a $\mathbb{Z}$-linear combination of the periodic vector fields generating the torus action. If s is only $\mathbb{Q}$-linear, we can rescale it to clear denominators and get a $\mathbb{Z}$-linear map, so the symplectic cut can still be made to work. If s is not $\mathbb{Q}$-linear then the subgroup of T^n generated by the flow of V_H is not closed. If we persist in taking the quotient, the result will likely fail to be Hausdorff.   □

Exercise 4.42 (Theorem 4.22) *Any convex rational polytope Δ occurs as the moment image of a toric Hamiltonian system (on a possibly singular space).*

Proof Start with the Hamiltonian system $\mu \colon T^*T^n \to \mathbb{R}^n$ given in canonical coordinates by $(p, q) \to p$. The moment image is the whole of $\mathbb{R}^n$. The polytope Δ is an intersection of a collection of half-spaces $s_j(x) \geq c_j$ where $s_1, \ldots, s_m$

are $\mathbb{Q}$-linear maps and $c_1, \ldots, c_m$ are real numbers. Take the symplectic cut of T^*T^n by $s_1 \circ \mu$ at level c_1. Then take the symplectic cut of the result by $s_2 \circ \mu$ at level c_2, and continue. Each time you cut, the moment image is intersected with another half-space. The final result is a Hamiltonian system whose moment image is Δ. The total space will have singularities if Δ is not Delzant. This will happen when we quotient by a non-free circle action. $\square$

Exercise 4.43 (Example 4.26) *There is a blow-up of $\mathbb{CP}^1 \times \mathbb{CP}^1$ in two disjoint symplectic balls which is symplectomorphic to a blow-up of $\mathbb{CP}^2$ in three disjoint balls.*

Proof Blow up the shaded balls by symplectic cut. This has the effect of truncating the moment polygons to obtain the moment hexagon of the common blow-up.

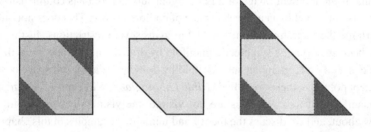

$\square$

Exercise 4.44 (Example 4.27) *Show that the common blow-up from Example 4.26 arises as a symplectic reduction of $\mathbb{CP}^1 \times \mathbb{CP}^1 \times \mathbb{CP}^1$.*

Proof Make the cut as shown.

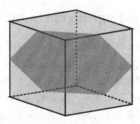

This gives a hexagonal moment polygon which is $\mathbb{Z}$-affine equivalent to the previous one (e.g. via projection to the xy-plane). $\square$

5

Visible Lagrangian Submanifolds

We will now study Lagrangian submanifolds of toric manifolds. It will turn out that if the moment image of a Lagrangian submanifold has codimension k then it is contained in an affine subspace of codimension k. This does not quite determine the Lagrangian completely, but gives severe restrictions. Just as we have been specifying a symplectic manifold by drawing a polytope, we will be able to specify a Lagrangian submanifold by drawing an affine subspace of the moment polytope; these are called *visible Lagrangians*. While most Lagrangian submanifolds of toric varieties are not visible, the visible ones are useful to know about, and we discuss the theory and numerous examples in this chapter. In Appendix H, we will see a more versatile construction due to Mikhalkin and Matessi, which assigns a *tropical Lagrangian* to a *tropical curve* in the polytope.

5.1 Visible Lagrangian Submanifolds

Theorem 5.1 *Consider the integrable Hamiltonian system $H : \mathbb{R}^n \times T^n \to \mathbb{R}^n$, $H(p, q) = p$ where $q_1, \ldots, q_n$ are taken modulo 2π and the symplectic form is $\sum dp_i \wedge dq_i$. Let $L \subseteq \mathbb{R}^n \times T^n$ be a Lagrangian submanifold. Suppose that $H|_L : L \to \mathbb{R}^n$ factors as $H|_L = f \circ g$, where $g : L \to K$ is a bundle over a k-dimensional manifold K, $k < n$, and $f : K \to \mathbb{R}^n$ is an embedding. Then K is an affine linear subspace of $\mathbb{R}^n$ which is rational with respect to the lattice $(2\pi\mathbb{Z})^n$.*

Definition 5.2 We call Lagrangian submanifolds which project in this way *visible*.

Remark 5.3   Theorem 5.1 was first observed when $n = 2$ and $\dim(K) = 1$ by Symington [106, Corollary 7.9].

Proof Let $s = (s_1, \ldots, s_k)$ be local coordinates on K and $t = (t_{k+1}, \ldots, t_n)$ be local coordinates on the fibre of g. By assumption, the inclusion of L into $\mathbb{R}^n \times T^n$ has the form $(s, t) \mapsto (p(s), q(s, t))$ for some functions p, q. The vectors ∂_{s_i} and ∂_{t_j} push-forward to $(\partial_{s_i} p, \partial_{s_i} q)$ and $(0, \partial_{t_j} q)$. The Lagrangian condition on L is equivalent to $\partial_{s_i} p \cdot \partial_{t_j} q = 0$ and $\partial_{s_i} p \cdot \partial_{s_j} q = \partial_{s_j} p \cdot \partial_{s_i} q$ for all i, j. The first of these conditions implies that the tangent space of the fibre of g is orthogonal[1] to the k-dimensional subspace $f_*(TK)$ spanned by $\partial_{s_1} p, \ldots, \partial_{s_k} p$. Since the tangent space of the fibre of g is $(n - k)$-dimensional, it must be precisely $f_*(TK)^\perp$; in other words, for each $s \in K$, the fibre of g over s is an integral submanifold of the distribution on T^n given by $f_*(TK)^\perp$. This distribution has an integral submanifold if and only if $f_*(TK)$ is a rational subspace with respect to the lattice $(2\pi\mathbb{Z})^n$. Since $f_*(TK)$ varies smoothly in s, and must always be rational, it is necessarily constant. Therefore $f(K)$ is a rational affine subspace. □

Remark 5.4   As a consequence of the proof, we see that if the visible Lagrangian projects to an affine subspace K in the p-plane then its fibre in the q-torus above a point in K is a translate of the subtorus $K^\perp/(K^\perp \cap (2\pi\mathbb{Z})^n)$.

Example 5.5   Suppose $n = 2$ and K is the p_1-axis. Then $L \cap \{(p_1, 0)\}$ is a circle $\{(q_1, \theta) : \theta \in [0, 2\pi]\}$ for some fixed q_1. For example, L could be the cylinder $\{p_2 = 0, \quad q_1 = 0\}$.

Remark 5.6   Note that the dependence of q_i on the coordinates s_j can be nontrivial.

Example 5.7   Let (p_1, p_2, q_1, q_2) be coordinates on $X = \mathbb{R}^2 \times T^2$ with symplectic form $\sum dp_i \wedge dq_i$. The Lagrangian embedding $i \colon \mathbb{R} \times S^1 \to X$, $i(s, t) = (s, 0, 0, t)$ is visible for the projection $(p, q) \mapsto p$. The Lagrangian torus $j \colon S^1 \times S^1 \to X$, $j(s, t) = (\sin s, 0, s, t)$ is also visible,[2] and it projects to the line segment $[-1, 1] \times \{0\}$ (the preimage of each point in $(-1, 1) \times \{0\}$ is a pair of circles).

Remark 5.8   Apart from giving a useful way to visualise and construct Lagrangian submanifolds, this theorem also gives us a way to figure out the integral affine structure on the base of a Lagrangian torus fibration if we don't already know it. If we can find a Lagrangian submanifold whose image under our Hamiltonian system is a submanifold $K \subseteq \mathbb{R}^n$ then we know that the image of K under action coordinates is supposed to be affine linear. We will use this observation in the proof of Lemma 7.2 later.

---

[1] Orthogonal with respect to the Euclidean metric on $\mathbb{R}^n$.

[2] Technically, it is not visible itself because the projection map is not a bundle, rather it is a union of two visible cylinders. We will tolerate this and related abuses of terminology.

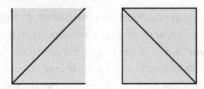

Figure 5.1 Left: A visible Lagrangian disc (Example 5.9). Right: The antidiagonal sphere in $S^2 \times S^2$ (Example 5.10) is a visible Lagrangian living over the antidiagonal in the square.

5.2 Hitting a Vertex

Suppose now that we have a Hamiltonian torus action (and toric critical points) with moment map $\mu\colon X \to \mathbb{R}^n$ and address the question of what visible Lagrangian surfaces look like when the affine linear subspace $\mu(L)$ intersects the boundary strata of the moment polytope. For simplicity, we will focus on the case $\dim X = 4$, $\dim \mu(L) = 1$.

Example 5.9 Consider the Lagrangian plane $L := \{(z, \bar{z}) \ : \ z \in \mathbb{C}\} \subseteq \mathbb{C}^2$. The projection $\mu(L)$ is the diagonal ray $\{(t, t) \ : \ t \in [0, \infty)\} \subseteq \mathbb{R}^2$, so L is a visible Lagrangian surface. See Figure 5.1.

Example 5.10 Consider the Lagrangian *antidiagonal sphere*

$$\bar{\Delta} := \{((x, y, z), (-x, -y, -z)) \in S^2 \times S^2 \ : (x, y, z) \in S^2\}.$$

Here, we have equipped $S^2 \times S^2$ with the equal-area symplectic form from Example 3.13. The moment map is $\mu((x_1, y_1, z_1), (x_2, y_2, z_2)) = (z_1, z_2)$ so the projection of the antidiagonal sphere along μ is the antidiagonal line $\{(z, -z) \in [-1, 1]^2 \ : \ z \in [-1, 1]\}$ (see Figure 5.1). This is therefore a visible Lagrangian whose projection hits two vertices, where it is locally modelled on Example 5.9.

Example 5.11 (Exercise 5.18) Fix $m, n \in \mathbb{Z}_{>0}$ with $\gcd(m, n) = 1$. Consider the ray $\{(mt, nt) \ : \ t \in [0, \infty)\}$ in the nonnegative quadrant (Figure 5.2). Above this ray is a visible Lagrangian which I will call a *Schoen–Wolfson cone*,[3] given parametrically by

$$(s, t) \mapsto \frac{1}{\sqrt{m+n}} \left(t\sqrt{m}e^{is\sqrt{n/m}}, it\sqrt{n}e^{-is\sqrt{m/n}}\right), \quad s \in [0, 2\pi\sqrt{mn}], \ t \in [0, \infty).$$

---

[3] Schoen and Wolfson [92, Theorem 7.1] showed that these are the only Lagrangian cones in $\mathbb{C}^2$ which are Hamiltonian stationary (i.e. critical points of the volume functional restricted to Hamiltonian deformations).

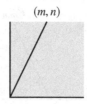

Figure 5.2 Moment image of a Schoen–Wolfson cone.

This cone is singular at the origin unless $m = n = 1$.

Remark 5.12   Modulo the freedom discussed in Remark 5.6 and Example 5.7, this exhausts all possible local models for visible Lagrangians living over a line which hits the corner of a Delzant moment polygon.

Example 5.13   If our moment polygon is a rectangle with sidelengths m and n (positive integers) then we get a symplectic form on $S^2 \times S^2$ which gives the factors symplectic area m and n respectively, so the symplectic form lives in the class $(m, n) \in H^2(S^2 \times S^2)$. The diagonal line joining opposite corners of the moment rectangle is the projection of a visible Lagrangian sphere L with two Schoen–Wolfson singular points with parameters (m, n). The homology class of this Lagrangian is $(n, -m)$ (you can see this by intersecting with spheres in the classes $[S^2 \times \{p\}]$ and $[\{p\} \times S^2]$), which has symplectic area 0. Indeed, the homology class $(n, -m)$ can only contain a Lagrangian representative when $[\omega]$ is a multiple of $(m, n) \in H^2(S^2 \times S^2)$.

5.3 Hitting an Edge

Example 5.14   We now consider visible Lagrangians whose projection hits an edge. For a local model, we take $X = \mathbb{R} \times S^1 \times \mathbb{C}$, with coordinates $(p, q, z = x+iy)$ ($q \in \mathbb{R}/2\pi\mathbb{Z}$) and symplectic form $dp \wedge dq + dx \wedge dy$. The image of the moment map $\mu \colon X \to \mathbb{R}^2$, $\mu(p, q, z) = \left(p, \frac{1}{2}|z|^2\right)$ is the closed upper half-plane $\{(x_1, x_2) \in \mathbb{R}^2 \;:\; x_2 \geq 0\}$. Consider the ray $R_{m,n} = \{(ms, ns) \;:\; s \geq 0\}$. The following map is a Lagrangian immersion of the cylinder

$$i(s, t) = \left(ms, -nt, \sqrt{2ns}\,e^{imt}\right), \qquad (s, t) \in [0, \infty) \times S^1$$

whose projection along μ is the ray $R_{m,n}$. This immersion is an embedding away from $s = 0$, but it is n-to-1 along the circle $s = 0$ (the points $\left(0, t + \frac{2\pi k}{n}\right)$, $k = 0, \ldots, n - 1$, all project to $(0, t \mod 2\pi, 0)$).

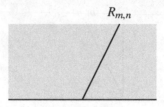

The image of the immersion is a Lagrangian which looks like a collection of n flanges meeting along a circle, twisting as they move around the circle so that the link of the circle is an (m, n)-torus knot (see Figure 5.3). For example, when $m = 1$, $n = 2$, this is a Möbius strip. For $n \geq 3$ it is not a submanifold. We call the image of the immersion a *Lagrangian (n, m)-pinwheel core*.

Any integral affine transformation preserving the upper half-plane and fixing the origin acts on the set of rays $R_{m,n}$. These transformations are precisely the affine shears $\begin{pmatrix} 1 & 0 \\ k & 1 \end{pmatrix}$, which allow us to change m by any multiple of n, so we can always assume $m \in \{0, \ldots, n - 1\}$.

Again, modulo the freedom discussed in Remark 5.6 and Example 5.7, these local models exhaust the visible Lagrangians intersecting an edge of a moment polygon.

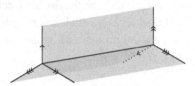

Figure 5.3 A pinwheel core with $n = 3$ flanges: lines with arrows should be identified in pairs.

Example 5.15 Consider the Lagrangian $\mathbb{RP}^2$ which is the closure of the visible disc $\{[z : \bar{z} : 1] : z \in \mathbb{C}\} \subseteq \mathbb{CP}^2$. This projects to the diagonal bisector in the moment triangle (see Figure 5.4(a)). If we use the integral affine transformation $\begin{pmatrix} -1 & -1 \\ 0 & -1 \end{pmatrix}$ to make the slanted edge of the triangle horizontal then the projection of the visible Lagrangian ends up pointing in the $\pm(1, 2)$-direction (Figure 5.4(b)), so comparison with Example 5.14 shows that the disc is capped off with a Möbius strip to give an $\mathbb{RP}^2$.

Example 5.16 (Exercise 5.19) Consider the square with vertices at $(-2, -2)$, $(-2, 2)$, $(2, -2)$, $(2, 2)$. There is a smooth, closed visible Lagrangian surface

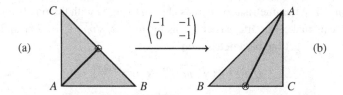

(a)   (b)

Figure 5.4 (a) Visible Lagrangian $\mathbb{RP}^2$ in $\mathbb{CP}^2$. (b) The same picture after an integral affine transformation shows the line pointing in the $(1, 2)$-direction, so there is a $(2, 1)$-pinwheel core (Möbius strip) near the point marked $\otimes$.

L in the corresponding toric variety, living over the line segment connecting $(-1, -2)$ to $(1, 2)$. To which topological surface is L homeomorphic?

Here is a higher-dimensional example.

Example 5.17 (Exercise 5.20)   Consider the symplectic manifold $\mathbb{CP}^1 \times \mathbb{C}^2$ with the symplectic form $pr_1^*\omega_{\mathbb{CP}^1} + pr_2^*\omega_{\mathbb{C}^2}$ (here, pr_k denotes the projection to the kth factor, $\omega_{\mathbb{CP}^1}$ is the Fubini–Study form on $\mathbb{CP}^1$ normalised so that $\frac{1}{2\pi} \int_{\mathbb{CP}^1} \omega_{\mathbb{CP}^1} = 1$, and $\omega_{\mathbb{C}^2}$ is the standard symplectic form). Sketch the moment image for the T^3-action coming from the standard torus actions on each factor. Check that the 3-sphere $\{([-\bar{z}_2 : \bar{z}_1], z_1, z_2) \ : \ |z_1|^2 + |z_2|^2 = 2\} \subseteq \mathbb{CP}^1 \times \mathbb{C}^2$ is Lagrangian and sketch its projection under the moment map.

5.4 Solutions to Inline Exercises

Exercise 5.18 (Example 5.11)   *Fix* $m, n \in \mathbb{Z}_{>0}$ *with* $\gcd(m, n) = 1$. *Verify that the Schoen–Wolfson cone*

$$(s, t) \mapsto \frac{1}{\sqrt{m+n}} \left(t\sqrt{m}e^{is\sqrt{n/m}}, it\sqrt{n}e^{-is\sqrt{m/n}} \right), \quad s \in [0, 2\pi\sqrt{mn}], \ t \in [0, \infty)$$

is Lagrangian where that makes sense (i.e. away from the cone point) and that its projection under the moment map is the ray $\{(mt, nt) \ : \ t \in [0, \infty)\}$.

Solution   The parametrisation here is chosen to agree with the one from [92], but we can make our life easier by using $\theta = s/\sqrt{mn} \in [0, 2\pi]$ and $r = t^2/2(m + n)$ to get the parametrisation

$$(r, \theta) \mapsto (\sqrt{2mr}\,e^{in\theta}, \sqrt{2nr}\,e^{-im\theta}).$$

Applying the moment map $(|z_1|^2/2, |z_2|^2/2)$ gives us $\{(mr, nr) : r \geq 0\}$, so the Lagrangian projects to the correct ray. The fibre of the Lagrangian over (mr, nr) is $\{(\sqrt{2mr}\,e^{in\theta}, \sqrt{2nr}\,e^{-im\theta}) : \theta \in [0, 2\pi]\}$. In the (θ_1, θ_2)-torus, this is a circle whose tangent line is $(n, -m)$, which is orthogonal to the ray in the base. Therefore, this is Lagrangian by Remark 5.4.   □

Exercise 5.19 (Example 5.16)   *Consider the square with vertices at* $(-2, -2)$, $(-2, 2)$, $(2, -2)$, $(2, 2)$. *There is a smooth, closed visible Lagrangian surface* L *in the corresponding toric variety, living over the line segment connecting* $(-1, -2)$ *to* $(1, 2)$. *To which topological surface is* L *homeomorphic?*

$(1, 2)$

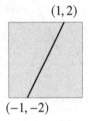

$(-1, -2)$

Solution   There are two Möbius strips where the projection of the visible Lagrangian meets the edge of the square. These are joined along their common boundary, which forms a Lagrangian Klein bottle.   □

Exercise 5.20 (Example 5.17)   *Consider the symplectic manifold* $\mathbb{CP}^1 \times \mathbb{C}^2$ *with the symplectic form* $pr_1^*\omega_{\mathbb{CP}^1} + pr_2^*\omega_{\mathbb{C}^2}$ *(here,* pr_k *denotes the projection to the* k*th factor,* $\omega_{\mathbb{CP}^1}$ *is the Fubini–Study form on* $\mathbb{CP}^1$ *normalised so that* $\frac{1}{2\pi}\int_{\mathbb{CP}^1}\omega_{\mathbb{CP}^1} = 1$ *and* $\omega_{\mathbb{C}^2}$ *is the standard symplectic form). Sketch the moment image for the* T^3*-action coming from the standard torus actions on each factor. Check that the 3-sphere* $L := \{([-\bar{z}_2 : \bar{z}_1], z_1, z_2) : |z_1|^2 + |z_2|^2 = 2\} \subseteq \mathbb{CP}^1 \times \mathbb{C}^2$ *is Lagrangian and sketch its projection under the moment map.*

Solution   The moment map is $\mu([a : b], z_1, z_2) = \left(\frac{|b|^2}{|a|^2 + |b|^2}, \frac{1}{2}|z_1|^2, \frac{1}{2}|z_2|^2\right)$, so its image is the noncompact polytope $\{(x, y, z) \in \mathbb{R}^3 : x \in [0, 1], y, z \geq 0\}$. To compute the moment image of L, we have

$$\mu([-\bar{z}_2 : \bar{z}_1], z_1, z_2) = \left(\frac{1}{2}|z_1|^2, \frac{1}{2}|z_1|^2, \frac{1}{2}(2 - |z_1|^2)\right),$$

where we used the fact that $|z_1|^2 + |z_2|^2 = 2$. As $|z_1|^2$ varies between 0 and 2 (again using the constraint $|z_1|^2 + |z_2|^2 = 2$), we get the straight line segment

$$t \mapsto (t/2, t/2, (2-t)/2), \qquad t \in [0, 2]$$

in the $y + z = 1$ plane, connecting $(0, 0, 1)$ to $(1, 1, 0)$.

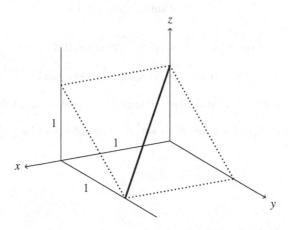

Figure 5.5 The moment image (slanted line) of the Lagrangian sphere in Exercise 5.20. The dotted rectangle is the plane $y + z = 1$ and numbers indicate affine lengths of edge-segments.

We now check that L is Lagrangian for the symplectic form $pr_1^*\omega_{\mathbb{CP}^1} + pr_2^*\omega_{\mathbb{C}^2}$. By definition of the Fubini–Study form, it suffices to lift the embedding $L \to \mathbb{CP}^1 \times \mathbb{C}^2$ to a map $L \to S^3(\sqrt{2}) \times \mathbb{C}^2$ where $S^3(\sqrt{2})$ is the sphere of radius $\sqrt{2}$ in $\mathbb{C}^2$, since $(\mathbb{CP}^1, \omega_{\mathbb{CP}^1})$ is obtained by the symplectic reduction $S^3(\sqrt{2})/S^1$. We choose the lift $\ell \colon (z_1, z_2) \mapsto (-\bar{z}_2, \bar{z}_1, z_1, z_2)$ restricted to $|z_1|^2 + |z_2|^2 = 2$.

To begin, we work on the solid torus $|z_1|^2 \leq 1$ inside $|z_1|^2 + |z_2|^2 = 2$: a similar argument holds on the complementary solid torus $|z_2|^2 \leq 1$. Pick coordinates r, θ, ϕ so that $z_1 = re^{i\theta}$ and $z_2 = \sqrt{2-r^2}e^{i\phi}$. We write out ℓ fully in these coordinates (separating real and imaginary parts):

$$\ell(r, \theta, \phi) = \Big(-\sqrt{2-r^2}\cos\phi, \sqrt{2-r^2}\sin\phi, r\cos\theta, r\sin\theta,$$
$$r\cos\theta, -r\sin\theta, \sqrt{2-r^2}\cos\phi, \sqrt{2-r^2}\sin\theta\Big)$$

This gives:

$$\ell_*\partial_r = \left(\frac{r\cos\phi}{\sqrt{2-r^2}}, -\frac{r\sin\phi}{\sqrt{2-r^2}}, \cos\theta, \sin\theta, \right.$$

$$\left. \cos\theta, -\sin\theta, -\frac{r\cos\phi}{\sqrt{2-r^2}}, -\frac{r\sin\phi}{\sqrt{2-r^2}} \right)$$

$$\ell_*\partial_\theta = (0, 0, -r\sin\theta, r\cos\theta,$$

$$-r\sin\theta, -r\cos\theta, 0, 0)$$

$$\ell_*\partial_\phi = \left(\sqrt{2-r^2}\sin\phi, \sqrt{2-r^2}\cos\phi, 0, 0, \right.$$

$$\left. 0, 0, -\sqrt{2-r^2}\sin\phi, \sqrt{2-r^2}\cos\phi \right),$$

from which one can check that all possible evaluations of ω vanish, for example

$$\omega(\ell_*\partial_r, \ell_*\partial_\phi) = r\cos^2\phi + r\sin^2\phi - r\cos^2\phi - r\sin^2\phi = 0. \qquad \square$$

6

Focus–Focus Singularities

So far, we have studied Lagrangian torus fibrations which have either no critical points or else toric critical points. In this chapter, we discuss another kind of critical point: the *focus–focus critical point*. Lagrangian torus fibrations whose critical points are either toric or focus–focus type are called *almost toric*. Allowing these critical points will drastically expand our zoo of examples, but also make our life more complicated. The main new features are that (a) the integral affine structure on the base of the fibration has nontrivial affine monodromy around the critical points, and (b) the integral affine base no longer uniquely determines the torus fibration.

6.1 Focus–Focus Critical Points

Example 6.1 (Standard focus–focus system)   Consider the following pair of Hamiltonians on $(\mathbb{R}^4, dp_1 \wedge dq_1 + dp_2 \wedge dq_2)$:

$$F_1 = -p_1 q_1 - p_2 q_2, \qquad F_2 = p_2 q_1 - p_1 q_2.$$

If we introduce complex coordinates[1] $p = p_1 + i p_2$, $q = q_1 + i q_2$ then $F := F_1 + i F_2 = -\bar{p} q$.

Lemma 6.2 (Exercise 6.18)   *The Hamiltonians F_1 and F_2 Poisson-commute. The Hamiltonian F_1 generates the $\mathbb{R}$-action $(p, q) \mapsto (e^t p, e^{-t} q)$. The Hamiltonian F_2 generates the circle action $(p, q) \mapsto (e^{it} p, e^{it} q)$.*

The orbits of the resulting $\mathbb{R} \times S^1$-action are:

- the origin (fixed point);

---

[1]   These complex coordinates are not supposed to be compatible with ω; indeed the p-plane and q-plane are both Lagrangian.

- the Lagrangian cylinders $P := \{(p, 0) \; : \; p \neq 0\}$ and $Q := \{(0, q) \; : \; q \neq 0\}$;
- and the Lagrangian cylinders $\{(p, q) \; : \; \bar{p}q = c\}$ for $c \in \mathbb{C} \setminus \{0\}$.

The diagram that follows represents the projection of $\mathbb{R}^4$ to $\mathbb{R}^2$ via

$$(p_1, p_2, q_1, q_2) \mapsto (|p|, |q|);$$

the projections of the $\phi_t^{F_1}$-flowlines are the hyperbolae; $\phi_t^{F_2}$-flowlines project to points. The Lagrangian cylinders P and Q are shown living over the axes, the fixed point is marked with a dot at the origin.

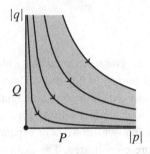

Definition 6.3 A *focus–focus chart* for an integrable Hamiltonian system $H : X \to \mathbb{R}^2$ is a pair of embeddings $E : U \to X$ and $e : V \to \mathbb{R}^2$ where:

- $U \subseteq \mathbb{R}^4$ is a neighbourhood of the origin and $E^* \omega = \sum dp_i \wedge dq_i$,
- $V = F(U)$, where F is the Hamiltonian system in Example 6.1,
- $H \circ E = e \circ F$.

We say that $H : X \to \mathbb{R}^2$ has a *focus–focus critical point* at $x \in X$ if there is a focus–focus chart (E, e) with $E(0) = x$.

Remark 6.4   This is not the standard definition of a focus–focus critical point: usually one specifies that H has a critical point at x and that the subspace of the space of quadratic forms spanned by the Hessians of the components H at x agrees with the corresponding subspace for F at 0. The fact that these two definitions are equivalent is a special case of *Eliasson's normal form theorem* for nondegenerate critical points of Hamiltonian systems. For a proof of this special case, see [15].

Lemma 6.5 ([112, Proposition 6.2])   *Let $H : X \to \mathbb{R}^2$ be an integrable Hamiltonian system with a focus–focus critical point x over the origin and no other critical points. The fibre $H^{-1}(0)$ is homeomorphic to a pinched torus.*

Proof   Recall that all our integrable systems are assumed to have compact, connected fibres. The fibre $H^{-1}(0)$ is a union of orbits $\Omega_0 = \{x\}, \Omega_1, \ldots, \Omega_k$

for the $\mathbb{R}^2$-action generated by $\boldsymbol{H}$. Since $\boldsymbol{H}^{-1}(0) \setminus \{x\}$ consists of regular points, the orbits $\Omega_1, \ldots, \Omega_k$ are two-dimensional submanifolds. By Theorem 1.40, there are three possible topologies of orbit: $\mathbb{R}^2$, $\mathbb{R} \times S^1$, and T^2. The third type cannot occur because it would give a connected component of $\boldsymbol{H}^{-1}(0)$ not containing x, but we assume our fibres are connected. In particular, any remaining orbits are noncompact.

Let Ω_P be the orbit containing the Lagrangian plane $E(P)$ (in the focus–focus chart) and Ω_Q be the orbit containing $E(Q)$. Note that it is possible that $\Omega_P = \Omega_Q$. Since the action of $\mathbb{R}^2$ on P (and on Q) has stabiliser $\mathbb{Z}$, these orbits are of the form $\mathbb{R} \times S^1$. Moreover, these are the only orbits containing $\{x\}$ in their closure (such an orbit must enter the focus–focus chart, where we can see that only Ω_P and Ω_Q contain x in their closure). There are two possibilities:

- $\Omega_P \neq \Omega_Q$. In this case, the union $\Omega_P \cup \{x\} \cup \Omega_Q$ would be noncompact. It is impossible to make $\boldsymbol{H}^{-1}(0)$ compact by adding further noncompact orbits, so because we assume H is proper, this possibility does not occur.
- $\Omega_P = \Omega_Q$. In this case, the cylinder Ω_P has both its ends attached to the point x, yielding a (compact) pinched torus. Adding further noncompact orbits contradicts compactness of $\boldsymbol{H}^{-1}(0)$, so there are no further orbits. □

The following figure shows a pinched torus fibre containing a focus–focus critical point. The fixed point is shown with a dot, and the $\phi_t^{H_2}$-flowlines are the short loops going around the fibre; the $\phi_t^{H_1}$-flowlines are the longer orbits connecting the fixed point to itself.

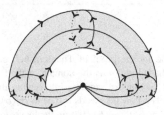

Remark 6.6   The same argument generalises to show that if $\boldsymbol{H}^{-1}(0)$ contains $m > 1$ focus–focus critical points then it will form a cycle of Lagrangian spheres, each intersecting the next transversely at a single focus–focus point (or, if $m = 2$, two spheres intersecting transversely at two points).

6.2 Action Coordinates

Let $\boldsymbol{H} \colon X \to \mathbb{R}^2$ be an integrable Hamiltonian system with a focus–focus critical point x over the origin and no other critical points. Let $E \colon U \to$

X, $e\colon V \to \mathbb{R}^2 = \mathbb{C}$ be a focus–focus chart centred at x. Recall that $F = F_1 + iF_2\colon U \to V$ denotes the model Hamiltonian from Example 6.1. Let $H_1 = F_1 \circ E^{-1}\colon E(U) \to \mathbb{R}$ and $H_2 = F_2 \circ E^{-1}\colon E(U) \to \mathbb{R}$. By shrinking U and V if necessary, assume that $V = \{\boldsymbol{b} \in \mathbb{R}^2 \colon |\boldsymbol{b}| < \epsilon\}$ for some $\epsilon > 0$; write $B := V \setminus \{0\}$ for the set of regular values of H. By Corollary 2.18, B inherits an integral affine structure, coming from action coordinates on the universal cover $\tilde{B}$. The next theorem identifies these action coordinates.

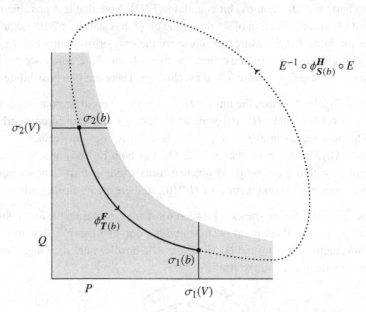

Figure 6.1 A schematic for the proof of Theorem 6.7, projected down to the $(|p|, |q|)$-plane. The shaded region is the domain of the focus–focus chart, and we see the Lagrangian sections σ_1 and σ_2 intersecting the Lagrangian planes P and Q respectively. The flow $\phi_{T(b)}^{F}$ sends $\sigma_2(b)$ to $\sigma_1(b)$ for $b \neq 0$. The flow $E^{-1} \circ \phi_{S(b)}^{H} \circ E$ sends $\sigma_1(b)$ to $\sigma_2(b)$ (exiting the focus–focus chart).

Theorem 6.7 (San Vũ Ngọc) *The action map $\tilde{B} \to \mathbb{R}^2$ has the form*

$$\left(\frac{1}{2\pi} \left(S(b) + b_2\theta - b_1(\log r - 1) \right), b_2 \right),$$

where $b = b_1 + ib_2 = re^{i\theta}$ is the local coordinate on B and $S(b)$ is a smooth function.

Proof The map $\sigma_1\colon V \to \mathbb{R}^4 = \mathbb{C}^2$, $\sigma_1(b) = (-1, b)$ is a Lagrangian section of F which intersects Q at $\sigma_1(0)$. Similarly $\sigma_2(b) = (-\bar{b}, 1)$ is a Lagrangian section which intersects P. See Figure 6.1.

For $b \neq 0$, we can use the Hamiltonians F_1 and F_2 *inside* our focus–focus chart to flow the point $\sigma_2(b) = (-\bar{b}, 1)$ until it hits $\sigma_1(b) = (-1, b)$ (see Figure 6.1). In other words, we can find functions $T_1(b)$ and $T_2(b)$ on $V \setminus \{0\}$ with:

$$\phi^{F_2}_{T_2(b)}\phi^{F_2}_{T_1(b)}(-\bar{b}, 1) = (-e^{T_1(b)+iT_2(b)}\bar{b}, e^{-T_1(b)+iT_2(b)}) = (-1, b), \tag{6.1}$$

namely

$$T_1(b) = -\ln|b|, \quad T_2(b) = \arg(b).$$

Claim: After possibly shrinking V, there exist smooth functions $S_1(b)$ and $S_2(b)$ defined on V such that

$$\phi^{H_2}_{S_2(b)}\phi^{H_1}_{S_1(b)}(E(\sigma_1(b))) = E(\sigma_2(b)) \tag{6.2}$$

for all $b \in V$ and such that

$$S_1 = \frac{\partial S}{\partial b_1}, \quad S_2 = \frac{\partial S}{\partial b_2} \tag{6.3}$$

for some smooth function S on V.

Let us see how the claim implies the theorem. Since $H_k = F_k \circ E^{-1}$, we can combine Equations (6.1) and (6.2) to get

$$\phi^{H}_{S+T}(E(\sigma_2(b)) = E(\sigma_2(b))$$

for all $b \in V$, so that $S + T := (S_1(b) + T_1(b), S_2(b) + T_2(b))$ is in the period lattice (see Figure 6.1). The period lattice is then spanned by these vectors and by $(0, 2\pi)$ (since H_2 already has period 2π). To find action coordinates (G_1, G_2), it suffices to solve

$$\begin{pmatrix} \frac{\partial G_1}{\partial b_1} & \frac{\partial G_1}{\partial b_2} \\ \frac{\partial G_2}{\partial b_1} & \frac{\partial G_2}{\partial b_2} \end{pmatrix} = \begin{pmatrix} \frac{1}{2\pi}(S_1(b) - \ln|b|) & \frac{1}{2\pi}(S_2(b) + arg(b)) \\ 0 & 1 \end{pmatrix}.$$

We can take $G_1(b) = \frac{1}{2\pi}(S + b_2\theta - b_1(\log r - 1))$ and $G_2(b) = b_2$ where $\theta = \arg(b)$ and $r = |b|$. This proves the theorem.

We now prove the claim. In the proof of Lemma 6.5, we saw that the branch $E(P)$ is part of the same $\mathbb{R}^2$-orbit as the branch $E(Q)$. Therefore, if we flow $E(\sigma_1(0))$ for using H_1 for some duration s_1, we will reach a point in $E(Q)$ at the same radius as $\sigma_2(0)$. Further flowing using H_2 for some time s_2, which preserves the radius in the Q-plane, we can ensure that

$$\phi^{H_2}_{s_2}\phi^{H_1}_{s_1}(\sigma_1(0)) = \sigma_2(0).$$

After possibly shrinking V, we get local Liouville coordinates near $\sigma_2(0)$ using the Lagrangian section σ_1:

$$\Psi(b, u) := \phi_{u_2}^{H_2} \phi_{u_1}^{H_1}(\sigma_1(b)).$$

The domain of Ψ is $V \times I$ where I is a neighbourhood of (s_1, s_2) in $\mathbb{R}^2$. The preimage $L := \Psi^{-1}(\sigma_2(V))$ is the Lagrangian submanifold of $V \times I$ given by

$$L = \{(b, u) \in V \times I \ : \ \phi_{u_2}^{H_2} \phi_{u_1}^{H_1}(1, -b) = (-\bar{b}, 1)\}.$$

We pick the unique component of L containing $(0, (s_1, s_2))$. This can be written as the graph of a function $b \mapsto (S_1(b), S_2(b))$. All that remains is to prove the following claim, which we leave as an exercise (Exercise 6.19). The graph $\{(b, (S_1(b), S_2(b))) \ : \ b \in V\}$ is Lagrangian if and only if $\partial S_1/\partial b_2 = \partial S_2/\partial b_1$, which holds if and only if $S_1 = \partial S/\partial b_1$ and $S_2 = \partial S/\partial b_2$ for some function S.

$\square$

Remark 6.8   In fact, any such S arises, as we will show in the next section. Moreover, Vũ Ngọc [111] showed[2] that the germ of S near the origin is unchanged by any fibred symplectomorphism of the system, and that this germ determines the (germ of the) system up to fibred symplectomorphism in a neighbourhood of the nodal fibre. We will write $(S)^\infty$ for the Vũ Ngọc invariant of a focus–focus critical point.

Remark 6.9   The action map has a well-defined limit point as $r \to 0$. We call this limit point the *base-node* of the focus–focus critical point.

6.3 Monodromy

We briefly recall the notion of affine monodromy introduced in Definition 2.19. Let $f \colon X \to B$ be a regular Lagrangian fibration, let $\tilde{B} \to B$ be the universal cover of the base of a Lagrangian fibration, and let $I \colon \tilde{B} \to \mathbb{R}^n$ be the developing map for the integral affine structure. Given an element $g \in \pi_1(B)$, we get a deck transformation $\tilde{b} \mapsto \tilde{b}g$ of the universal cover, and $I(\tilde{b}g) = I(\tilde{b})M(g)$ for some matrix $M(g) \in SL(n, \mathbb{Z})$.

Example 6.10   Let H be an integrable Hamiltonian system with a single focus–focus fibre, and let f be the restriction of H to the complement of the focus–focus fibre. In this case, $B = \mathbb{R}^2 \setminus \{0\}$ with polar coordinates r, θ. The universal cover $\tilde{B}$ is obtained by treating θ as a real-valued (instead of periodic

---

[2] There is a subtlety here: the germ of S can depend on the choice of focus–focus chart. This is a finite ambiguity and is discussed in [95, Section 4.3]: the actual Vũ Ngọc invariant is an equivalence class of germs under an action of the Klein 4-group.

angular) coordinate. As a corollary of 6.7, we get the developing map for the integral affine structure:

$$I(r, \theta) = \left(\frac{1}{2\pi} \left(S(b) + b_2\theta - b_1(\log r - 1) \right), b_2 \right),$$

where $(b_1, b_2) = (r\cos\theta, r\sin\theta)$. We have $\pi_1(B) = \mathbb{Z}$ and $n \in \mathbb{Z}$ acts on $\tilde{B}$ by $(r, \theta) \mapsto (r, \theta + 2\pi n)$.

Lemma 6.11 (Exercise 6.20)   *The affine monodromy for* $n \in \pi_1(B)$ *is* $M(n) = \begin{pmatrix} 1 & 0 \\ n & 1 \end{pmatrix}$.

This means that if you 'go around the loop' in B, the action map changes by this shear matrix. Figures 6.2 and 6.3 illustrate this by plotting the image under I of some different choices of fundamental domain for the covering map $\tilde{B} \to B$ (for the choice $S \equiv 0$). We include the images under the action map of contours of constant r (encircling the origin) and constant θ (pointing roughly radially outward).

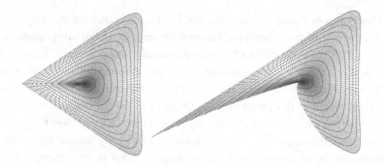

Figure 6.2 Left: The image of the fundamental domain $\{\theta \in [-\pi, \pi)\}$. This 'closes up' in the sense that $\theta = -\pi$ and $\theta = \pi$ map to the same line. This is because $\theta = \pi$ is an eigenline of the monodromy matrix. Right: The image of the fundamental domain $\{\theta \in [-5\pi/7, 9\pi/7)\}$. Although this plot does not 'close up', the image of the radius $\theta = -5\pi/7$ and the image of the radius $\theta = 9\pi/7$ are related by the monodromy matrix.

Remark 6.12   We think of B as being obtained from $\tilde{B}$ as follows. Fix a number $\theta_0 \in \mathbb{R}$, take the fundamental domain $(0, \infty) \times [\theta_0, \theta_0 + 2\pi] \subseteq \tilde{B}$ and identify points (r, θ_0) with $(r, \theta_0 + 2\pi)$. In other words, we make a 'branch cut' at $\theta = \theta_0$. We pull back the integral affine structure on $\mathbb{R}^2$ along the developing map I to get an integral affine structure on $\tilde{B}$, and this descends to B as in Corollary 2.18. When we make the identification $(r, \theta_0) \sim (r, \theta_0 + 2\pi)$, we need to specify how to

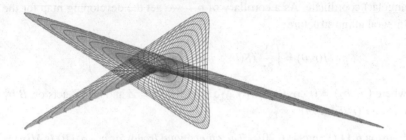

Figure 6.3 In the third figure, we see the image of two fundamental domains $\{\theta \in [-5\pi/2, 3\pi/2)\}$, related to one another by the action of the monodromy matrix. Anyone who has compulsively traced out the spiral of a raffia mat cannot fail to be moved by this image.

identify the integral affine structures. We use $I(r, \theta_0 + 2\pi) = I(r, \theta_0)M(1)$, that is $I(r, \theta_0 + 2\pi)M(-1) = I(r, \theta_0)$. In other words, *when we cross the branch cut anticlockwise (direction of increasing θ), we apply the transformation $M(-1)$ to tangent vectors.*

Observe from Figures 6.2 and 6.3 that if we use a branch cut $\theta = 0$ or $\theta = \pi$ (parallel to the eigenline of the affine monodromy) to cut out our fundamental domain in $\tilde{B}$ then the image of this fundamental domain under I 'closes up', that is, it is surjective onto a punctured neighbourhood of the origin. If we use a different branch cut then the image of the fundamental domain under I will avoid a segment of this punctured neighbourhood. For this reason, we usually work with branch cuts parallel to the eigenline of the affine monodromy.

Note that the action map from Theorem 6.7 is only unique up to post-composition by an integral affine transformation. That is, by Lemma 1.41, if we post-compose I by an integral affine transformation[3] $\alpha(b) = bA + C$ (for some $A \in GL(2, \mathbb{Z})$ and $C \in \mathbb{R}^n$) then we do not change the period lattice, so we get an alternative set of action coordinates.

Lemma 6.13 (Exercise 6.21) *If we use the action coordinates IA for some $A \in SL(2, \mathbb{Z})$ then the clockwise affine monodromy is given by $A^{-1}M(1)A$ and the line of eigenvectors points in the $(1, 0)A$-direction. More precisely, if $(1, 0)A = (p, q)$ for some pair of coprime integers p, q and $\det(A) = 1$ then*
$$A^{-1}MA = \begin{pmatrix} 1 - pq & -q^2 \\ p^2 & 1 + pq \end{pmatrix}.$$

Remark 6.14 Remember that this matrix is acting *on the right*; if you want

---

[3] Here b is a row vector and A is acting on the right.

to think of your action coordinates as column vectors, you need to take the transpose matrix. Remember also that this is the *clockwise* monodromy: you apply its inverse to tangent vectors when you cross the branch cut anticlockwise.

6.4 Visible Lagrangians

The following visible Lagrangian disc will play an important role in our future analysis of focus–focus systems.

Lemma 6.15 *Let $H: X \to \mathbb{R}^2$ be an integrable Hamiltonian system with a focus–focus critical point at $x \in X$, let B be the set of regular values and $\tilde{B}$ its universal cover, and let $I: \tilde{B} \to \mathbb{R}^2$ be the developing map for the integral affine structure on B coming from action coordinates. Let $b \in \mathbb{R}^2$ be the base-node associated to the focus–focus critical point at x. Suppose that ℓ is a straight ray in $\mathbb{R}^2$ emanating from b pointing in an eigendirection for the affine monodromy around the critical value. Then there is a visible Lagrangian disc living over ℓ.*

Proof In the focus–focus chart we can simply use the Lagrangian disc $q = p$, which satisfies $F(p, p) = -\bar{p}p$, so this lives over the negative b_1-axis ($b_2 = 0$). By Theorem 6.7, the image of this under I is still the negative b_1-axis, which is an eigenray of the affine monodromy. □

Definition 6.16 By analogy with a similar (but slightly different[4]) situation in Picard–Lefschetz theory, this visible Lagrangian disc is called the *vanishing thimble* for the focus–focus critical point, and its intersection with any fibre over the ray ℓ is a loop in the fibre called the *vanishing cycle*.

6.5 Model Neighbourhoods

We now present a construction due to Vũ Ngọc which, given a function $S: \mathbb{R}^2 \to \mathbb{R}$, produces a Hamiltonian system $H_S: X_S \to \mathbb{R}^2$ with a focus–focus critical point whose Vũ Ngọc invariant is $(S)^\infty$. We will write $S_i = \frac{\partial S}{\partial b_i}, i = 1, 2$.

Take the subset $X := \{(p, q) \in \mathbb{R}^4 : |\bar{p}q| < \epsilon\}$ equipped with the Hamiltonian system F from Example 6.1. We will construct two Liouville coordinate systems on different regions of this space.

Recall the Lagrangian sections $\sigma_1(b) = (-1, b)$ and $\sigma_2(b) = (-\bar{b}, 1)$. We construct a third Lagrangian section $\sigma_3(b) = (-e^{S_1(b)+iS_2(b)}, e^{-S_1(b)+iS_2(b)}b) =$

---

[4] In Picard–Lefschetz theory, we have a *holomorphic* fibration instead of a Lagrangian fibration, but the thimble is still a Lagrangian disc.

$\phi_S^H(\sigma_1(b))$. We can use these Lagrangian sections to construct Liouville coordinates

$$\Psi_2(b,t) = \phi_t^F(\sigma_2(b))$$

and

$$\Psi_3(b,t) = \phi_t^F(\sigma_3(b))$$

with $t_1 \in [0, \delta)$ and $t_2 \in [0, 2\pi)$.

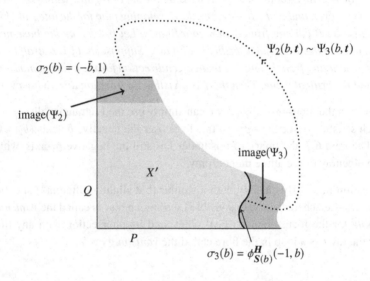

Figure 6.4 Construction of a Vũ Ngọc model with invariant S. The subset X' is the entire shaded region; the images of the Liouville coordinates $\Psi_2(b,t) = \phi_t^F(\sigma_2(b))$ and $\Psi_3(b,t) = \phi_t^F(\sigma_3(b))$ are shaded darker. The quotient X_S identifies these two darkly shaded regions.

Let $X' = \{(p,q) \in \mathbb{R}^4 : |\bar{p}q| < \epsilon, |q| \leq 1, |p| \leq e^{S_1(-\bar{p}q)}\}$ and let X_S be the quotient $X_S := X'/\sim$, where $\sim$ identifies $\Psi_2(b,t) \sim \Psi_3(b,t)$ (see Figure 6.4). Since the domains of Ψ_2 and Ψ_3 are identical and since Ψ_2 and Ψ_3 are symplectomorphisms, the symplectic form on X descends to this quotient. By construction, the map $H: X \to \mathbb{R}^2$, $H(p,q) = -\bar{p}q$ descends to the quotient and produces the Hamiltonian system H_S we want. Also by construction, the Vũ Ngọc invariant is $(S)^\infty$.

6.6 Symington's Theorem on Vũ Ngọc Models

We now present an argument of Symington [105, Lemma 3.6] which tells us that, although the Vũ Ngọc models $H_{S_0} \colon X_{S_0} \to \mathbb{R}^2$, $H_{S_1} \colon X_{S_1} \to \mathbb{R}^2$ with $(S_0)^\infty \neq (S_1)^\infty$ are not symplectomorphic via a fibred symplectomorphism, there is nonetheless a symplectomorphism $X_{S_0} \to X_{S_1}$ which is fibred outside a compact set. Because the fibred symplectomorphism type of the system depends only on the germ of S near the origin, we may assume that S_0 and S_1 coincide outside a small neighbourhood of the origin.

Theorem 6.17 (Symington) *Let $S_0 \colon \mathbb{R}^2 \to \mathbb{R}$ and $S_1 \colon \mathbb{R}^2 \to \mathbb{R}$ be smooth functions which coincide on the complement of a small disc D centred at the origin and let $H_{S_0} \colon X_{S_0} \to \mathbb{R}^2$ and $H_{S_1} \colon X_{S_1} \to \mathbb{R}^2$ be the corresponding Vũ Ngọc models. Then there is a symplectomorphism $\varphi \colon X_{S_0} \to X_{S_1}$ which restricts to a fibred symplectomorphism $H_{S_0}^{-1}(\mathbb{R}^2 \setminus D) \to H_{S_1}^{-1}(\mathbb{R}^2 \setminus D)$.*

Proof Pick a family S_t interpolating between S_0 and S_1 such that $S_t|_{\mathbb{R}^2 \setminus D} = S_0|_{\mathbb{R}^2 \setminus D}$. Consider the family of symplectic manifolds $X_t := X_{S_t}$; since the construction depends only on S_t, which is independent of t on the complement of D, the subsets $U_t := H_{S_t}^{-1}(\mathbb{R}^2 \setminus D)$ are fibred-symplectomorphic (via the identity map).

We extend this identification to a isotopy of diffeomorphisms $\varphi_t \colon X_0 \to X_t$ such that $\varphi_t|_{U_0} = \mathrm{id} \colon U_0 \to U_t$. For example, we could pick a connection on the family X_t which is trivial on $\bigcup_{t \in [0,1]} U_t = U_0 \times [0,1]$ and take φ_t to be the parallel transport of fibres.

Consider the family of symplectic forms $\omega_t = \varphi_t^* \omega_{S_t}$ on X_0. These satisfy $\frac{d\omega_t}{dt} = 0$ on U_0. The 2-form $\frac{d\omega_t}{dt}$ therefore determines a class in $H^2_{dR}(X_0, U_0)$. We will show that this class vanishes for all t; this will allow us to pick a family of 1-forms β_t such that $d\beta_t = d\omega_t/dt$ and $\beta_t|_{U_0} = 0$. By Moser's trick (see Appendix E), we then get diffeomorphisms $\phi_t \colon X_{S_0} \to X_{S_0}$, equal to the identity outside U_0, such that $\phi_t^* \varphi_t^* \omega_{S_t} = \omega_{S_0}$. The symplectomorphism we want is $\varphi := \varphi_1 \circ \phi_1 \colon X_{S_0} \to X_{S_1}$.

It remains to show that $\frac{d[\omega_t]}{dt} = 0 \in H^2_{dR}(X_0, U_0)$. Let $V = X_0 \setminus U_0$. We have $H^2_{dR}(X_0, U_0) = H^2_{dR}(V, \partial V)$ by excision, and $H^2_{dR}(V, \partial V) = H_2(V)$ by Poincaré–Lefschetz duality. Since V deformation-retracts onto the nodal fibre, we have $H^2_{dR}(V) = \mathbb{R}$. The cohomology class of a closed 2-form on X_0 which vanishes on U_0 can therefore be detected by its integral over a disc in V with boundary on ∂V which intersects the nodal fibre once transversely, for example a section of H_{S_t}. The construction of X_{S_t} furnishes it with an ω_{S_t}-Lagrangian section (i.e. $(-\bar{b}, 1)$); let us write σ_t for this section viewed via φ_t as a submanifold of X_0. Since φ_t is the identity outside U_0, we have $\sigma_t \cap U_0 = \sigma_0 \cap U_0$. Since σ_t

is ω_t-Lagrangian, we have $0 = \int_{\sigma_t} \omega_t$. Fix T and let $\Sigma_T(\boldsymbol{b}, t) := \sigma_t(\boldsymbol{b})$ be the isotopy of sections restricted to $t \in [0, T]$. Since $d\omega_T = 0$, Stokes's theorem[5] tells us that

$$0 = \int_{\Sigma} \Sigma^* d\omega_T = \int_{\sigma_T} \omega_T - \int_{\sigma_0} \omega_T.$$

Since σ_T is ω_T-Lagrangian, we get $\int_{\sigma_0} \omega_T = 0$. Therefore

$$\int_{\sigma_0} \frac{d\omega_t}{dt} = \frac{d}{dt} \int_{\sigma_0} \omega_t = 0,$$

so $\frac{d[\omega_t]}{dt} = 0 \in H^2(X_0, U_0)$.   □

6.7 Solutions to Inline Exercises

Exercise 6.18 (Lemma 6.2) *Verify that the Hamiltonians $F_1 = -p_1 q_1 - p_2 q_2$ and $F_2 = p_2 q_1 - p_1 q_2$ Poisson-commute, that F_1 generates the $\mathbb{R}$-action $(p, q) \mapsto (e^t p, e^{-t} q)$, and that F_2 generates the circle action $(p, q) \mapsto (e^{it} p, e^{it} q)$.*

Solution We have $-dF_1 = p_1 \, dq_1 + p_2 \, dq_2 + q_1 \, dp_1 + q_2 \, dp_2$ which equals $\iota_{(p_1, -q_1, p_2, -q_2)}(dp_1 \wedge dq_1 + dp_2 \wedge dq_2)$. Thus $V_{F_1} = (p_1, -q_1, p_2, -q_2)$ and the flow satisfies $\dot{p} = p$, $\dot{q} = -q$, which means $p(t) = e^t p(0)$ and $q(t) = e^t q(0)$. Similarly, we find $V_{F_2} = (-p_2, -q_2, p_1, q_1)$, whose flow satisfies $\dot{p} = ip$ and $\dot{q} = iq$ (recall that $p = p_1 + ip_2$ and $q = q_1 + iq_2$) and the flow is therefore $p(t) = e^{it} p(0)$ and $q(t) = e^{it} q(0)$. To see that $\{F_1, F_2\} = 0$, we compute

$$\{F_1, F_2\} = \omega(V_{F_1}, V_{F_2}) = -p_1 q_2 - q_1 p_2 + p_2 q_1 - q_2 p_1 = 0. \quad □$$

Exercise 6.19 (From the proof of Theorem 6.7) *Let V be a disc in $\mathbb{R}^2$. The graph $\{(b, (S_1(b), S_2(b))) : b \in V\}$ is Lagrangian if and only if $\partial S_1 / \partial b_2 = \partial S_2 / \partial b_1$, which holds if and only if $S_1 = \partial S / \partial b_1$ and $S_2 = \partial S / \partial b_2$ for some function S.*

Solution The tangent space to the graph is spanned by the vectors

$$(1, 0, \partial S_1 / \partial b_1, \partial S_2, \partial b_2),$$
$$(0, 1, \partial S_1 / \partial b_2, \partial S_2 / \partial b_2),$$

on which the symplectic form evaluates to

$$\partial S_2 / \partial b_1 - \partial S_1 / \partial b_2.$$

---

[5] There should be further boundary terms corresponding to the boundary of the section, but since the sections are all fixed over U_0 these contributions vanish.

The graph is Lagrangian if and only if this quantity vanishes. This is equivalent to the condition that the 1-form

$$S_1 db_1 + S_2 db_2$$

is closed. Since the disc V has zero de Rham cohomology in degree 1, this 1-form is closed if and only if it is exact, that is, if and only if there exists a function S with $\partial S / \partial b_k = S_k$ for $k = 1, 2$. $\qquad\square$

Exercise 6.20 (Lemma 6.11) *The affine monodromy for $n \in \pi_1(B)$ in Example 6.10 is $M(n) = \begin{pmatrix} 1 & 0 \\ n & 1 \end{pmatrix}$.*

Solution Since

$$I(r, \theta) = \left(\frac{1}{2\pi} (S(b) + b_2\theta - b_1(\log r - 1)), \ b_2 \right),$$

we have

$$I(r, \theta + 2\pi n) = \left(\frac{1}{2\pi} (S(b) + b_2(\theta + 2\pi n) - b_1(\log r - 1)), \ b_2 \right),$$

$$= \left(\frac{1}{2\pi} (S(b) + b_2\theta - b_1(\log r - 1)) + nb_2, \ b_2 \right)$$

$$= I(r, \theta) \begin{pmatrix} 1 & 0 \\ n & 1 \end{pmatrix}. \qquad\square$$

Exercise 6.21 (Lemma 6.13) *Let I be the developing map from Theorem 6.7. If we use instead the developing map IA for some $A \in SL(2, \mathbb{Z})$ then the clockwise affine monodromy is given by $A^{-1}M(1)A$ and the line of eigenvectors points in the $(1, 0)A$-direction. More precisely, if $(1, 0)A = (p, q)$ for some pair of coprime integers p, q and $\det(A) = \pm 1$ then*

$$A^{-1}M(1)A = \begin{pmatrix} 1 \mp pq & \mp q^2 \\ \pm p^2 & 1 \pm pq \end{pmatrix}.$$

Solution More generally, if $g \in \pi_1(B)$ then we have $I(\tilde{b}g)A = I(\tilde{b})M(g)A = I(\tilde{b})A(A^{-1}M(g)A)$, so the affine monodromy associated to the developing map IA is $A^{-1}M(g)A$. Since $v = (1, 0)$ is an eigenvector of $M(g)$, so that $vM(g) = v$, then $vAA^{-1}M(g)A = vM(g)A = vA$, so vA is an eigenvector of $A^{-1}M(g)A$.

If $(1, 0)A = (p, q)$ then $A = \begin{pmatrix} p & q \\ k & \ell \end{pmatrix}$ for some $k, \ell \in \mathbb{Z}$ with $p\ell - kq = \pm 1$. Therefore,

$$A^{-1} \begin{pmatrix} 1 & 0 \\ 1 & 1 \end{pmatrix} A = \pm \begin{pmatrix} \ell & -q \\ -k & p \end{pmatrix} \begin{pmatrix} p & q \\ p+k & q+\ell \end{pmatrix} = \begin{pmatrix} 1 \mp pq & \mp q^2 \\ \pm p^2 & 1 \pm pq \end{pmatrix}. \qquad\square$$

7

Examples of Focus–Focus Systems

We are now ready to introduce the notion of an *almost toric manifold*: a symplectic 4-manifold with a Lagrangian torus fibration whose critical points can be both toric and focus–focus type. Before developing the general theory in Chapter 8, we explore some examples.

7.1 The Auroux System

Like many people, I first learned of the following example from the wonderful expository article [7] on mirror symmetry for Fano varieties by Denis Auroux, where it serves to illustrate the wall-crossing phenomenon for discs.

Example 7.1 (Auroux system) Fix a real number $c > 0$. Consider the Hamiltonians $\boldsymbol{H} = (H_1, H_2) \colon \mathbb{C}^2 \to \mathbb{R}^2$ defined by $H_1(z_1, z_2) = |z_1 z_2 - c|^2$ and $H_2(z_1, z_2) = \frac{1}{2}\left(|z_1|^2 - |z_2|^2\right)$. The flow of H_2 is $\phi_t^{H_2}(z_1, z_2) = (e^{it} z_1, e^{-it} z_2)$. This shows that $\{H_1, H_2\} = 0$, because H_1 is constant along the flow of H_2 (see Lemma 1.16). The flow of H_1 is harder to compute. We can nonetheless understand the orbits of this system geometrically.

Consider the holomorphic map $\pi \colon \mathbb{C}^2 \to \mathbb{C}$, $\pi(z_1, z_2) = z_1 z_2$. This is a conic fibration: the fibres $\pi^{-1}(p)$ are smooth conics except for $\pi^{-1}(0)$ which is a singular conic (union of the z_1- and z_2-axes).

86

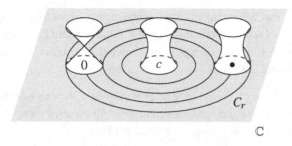

The Hamiltonian H_1 measures the squared distance in $\mathbb{C}$ from $z_1 z_2$ to some fixed point c. The level set $H_1^{-1}(r^2)$ is therefore the union of all conics living over a circle C_r of radius r centred at c (the concentric circles in the base of the figure). The restriction of H_2 to each conic can be visualised as a 'height function' whose level sets are circles as shown in the following diagrams. The level set $H^{-1}(b_1, b_2)$ is therefore the union of all circles of height b_2 in conics living over the circle $C_{\sqrt{b_1}}$. These level sets are clearly tori, except for the level set $H^{-1}(c^2, 0)$, which is a pinched torus.

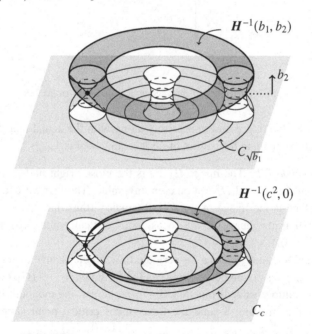

This system has a focus–focus critical point at $(0, 0)$. It also has toric critical points along the conic $z_1 z_2 = c$. Given a Hamiltonian system H and a critical point x of H, let $Q(H, x)$ denote the subspace of the space of quadratic forms spanned by the Hessians of the components of H. It is an exercise (Exercise

7.12) to check that, after a suitable symplectic change of coordinates, $Q(H, 0) = Q(F, 0)$, where H is the Auroux system and F is the standard focus–focus system from Example 6.1. This is enough to guarantee the existence of a focus–focus chart; see Remark 6.4.

Lemma 7.2 *Let B be the set of regular fibres of the Auroux system and $\tilde{B}$ its universal cover. There is a fundamental domain for the deck group action on $\tilde{B}$ whose image under action coordinates has the form*

$$\{(b_1, b_2) \ : \ 0 \le b_1 \le \phi(b_2)\} \setminus \{(b_1, 0) \ : \ b_1 \ge m\} \subseteq \mathbb{R}^2$$

for some function $\phi \colon \mathbb{R} \to (0, \infty)$ and some number $m > 0$ (see Figure 7.1). The affine monodromy, on crossing the branch cut $\{(b_1, 0) \ : \ b_1 \ge m\}$ clockwise, is $\begin{pmatrix} 1 & 0 \\ 1 & 1 \end{pmatrix}$.

Figure 7.1 The fundamental action domain from Lemma 7.2.

Remark 7.3   Finding ϕ and m precisely along with the actual map from the fundamental domain to this subset of $\mathbb{R}^2$ is a nontrivial task.

Proof of Lemma 7.2   The image $H(\mathbb{C}^2)$ is the closed right half-plane: H_1 is always nonnegative and H_2 can take on any value. The vertical boundary of the half-plane is the image of the toric boundary (the conic $z_1 z_2 = c$). The point $(c^2, 0)$ is the image of the focus–focus critical point $(0, 0)$ and $B = H(\mathbb{C}^2) \setminus \{(c^2, 0)\}$.

The Hamiltonian H_2 gives a 2π-periodic flow, so the change of coordinates of $\mathbb{R}^2$ which gives action coordinates has the form $(b_1, b_2) \mapsto (G_1(b_1, b_2), b_2)$ for some (multiply-valued) function G_1. In particular, the monodromy of the integral affine structure around the focus–focus critical point simply shifts amongst the branches of G_1, so has the form $\begin{pmatrix} 1 & 0 \\ 1 & 1 \end{pmatrix}$. We may make a branch cut along the line $R = \{(b_1, 0) \ : \ b_1 > c^2\}$ to get a simply-connected open set $U = B \setminus R$ and pick a fundamental domain $\tilde{U}$ lying over U in the universal cover $p \colon \tilde{B} \to B$.

We first compute the image $\{((G_1(0, b_2), b_2) \; : \; b_2 \in \mathbb{R}\}$ of the line $0 \times \mathbb{R}$ under the action coordinates. Since this is part of the toric boundary, Proposition 3.3 implies this image is a straight line S with rational slope. As observed in Lemma 6.15, there is a visible Lagrangian disc emanating from the focus–focus critical point and living over an eigenline of the affine monodromy. Actually, we can write the disc explicitly for the Auroux system: it is the Lagrangian disc $\{(z, \bar{z}) \; : \; |z|^2 \leq c\}$ with boundary on $z_1 z_2 = c$. This visible disc lives over the horizontal line segment $\{(b_1, 0) \; : \; b_1 \leq c^2\}$ under the map H and hence[1] over a horizontal line segment $\{((G_1(b_1, 0), 0) \; : \; b_1 \leq c^2\}$ in the image of action coordinates. This line segment connects S to the base-node $(G_1(c^2, 0), 0)$. Since this visible Lagrangian is a disc, not a pinwheel core, comparison with the local models from Example 5.14 shows that the line S must have slope $1/n$ for some integer n. In particular, post-composing action coordinates with an integral affine shear $\begin{pmatrix} 1 & 0 \\ -n & 1 \end{pmatrix}$, we get that S is vertical (we always have the freedom to post-compose our action coordinates with an integral affine transformation, thanks to Lemma 1.41). Now it is clear that the fundamental action domain has the required form, where $\phi(b_2) = \sup_{b_1 \in [0,\infty)} G_1(b_1, b_2)$ and $m = G_1(c^2, 0)$. □

7.2 Different Branch Cuts

We can always pick a different simply-connected domain $U \subseteq B$ to get well-defined action coordinates I, as we illustrated in Figure 6.2 in the previous chapter. The image of U will not in general 'close-up': unless we take a branch cut along the eigendirection of the affine monodromy, the boundary of $I(U)$ will be two branch cuts related by the affine monodromy.

To illustrate this, we will plot here some of the associated pictures for the Auroux system as the branch cut undergoes a full rotation. It is important to emphasise that all of these are integral affine bases for the *same* Hamiltonian system on the *same* manifold; they differ only in the choice of a fundamental domain for the covering space $\tilde{B} \to B$.

Remark 7.4   In some of these pictures, the toric boundary appears 'broken'. This is an artefact of the fact that it intersects the branch cut: the two segments of the toric boundary are related by the affine monodromy and therefore form one straight line in the integral affine structure. If you want to check this, the *anticlockwise* affine monodromy is $\begin{pmatrix} 1 & 0 \\ -1 & 1 \end{pmatrix}$, so, for example in the 9:00

[1]   See Remark 5.8.

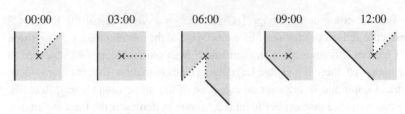

Figure 7.2 The Auroux system seen with different branch cuts; as we move from left to right in the figure, we see the branch cut rotate (from the 12-o'clock position) by 360 degrees. The final picture is related to the first by the affine monodromy.

diagram, the tangent vector $(0, -1)$ to the line above the branch cut gets sent to $(0, -1) \begin{pmatrix} 1 & 0 \\ -1 & 1 \end{pmatrix} = (1, -1)$ below the branch cut, which is tangent to the continuation of the boundary.

Remark 7.5   We can apply an integral affine transformation to any of these diagrams. Applying the matrix $\begin{pmatrix} 1 & 1 \\ 0 & 1 \end{pmatrix}$ to the 09:00 diagram in Figure 7.2 yields Figure 7.3 which will be important in the next chapter and which has anticlockwise affine monodromy $\begin{pmatrix} 2 & 1 \\ -1 & 0 \end{pmatrix}$ by Lemma 6.13. The importance of this example is that away from the branch cut, the integral affine manifold looks like the standard Delzant corner. We will see that this means we can always 'implant' this local Hamiltonian system whenever we have a polygon with a standard Delzant corner, an operation known as a *nodal trade*.

Figure 7.3 Another fundamental action domain for the Auroux system.

7.3 Smoothing A_n Singularities

Example 7.6   Let $P(z)$ be a polynomial of degree $n + 1$ with $n + 1$ distinct roots and $P(0) \neq 0$. Let $M_P = \{(z_1, z_2, z_3) \in \mathbb{C}^3 : z_1 z_2 + P(z_3) = 0\}$. If you allow P to vary, you get a family of such varieties; as P approaches the degenerate polynomial $P(z) = z^{n+1}$, the variety M_P develops a singularity called an A_n-

singularity. In other words, for generic P (with distinct roots) the variety M_P is the *Milnor fibre* (or *smoothing*[2]) of the A_n singularity (see Milnor's book [81] for more about Milnor fibres). Milnor fibres of singularities provide a rich class of symplectic manifolds which have been intensively studied.

Let $\pi\colon M_P \to \mathbb{C}$ be the conic fibration $\pi(z_1, z_2, z_3) = z_3$. By analogy with the Auroux system, we define

$$H(z_1, z_2, z_3) = \left(|z_3|^2, \frac{1}{2}\left(|z_1|^2 - |z_2|^2\right)\right).$$

Again, these Hamiltonians commute with one another, but only H_2 generates a circle action.

The subvariety $z_3 = 0$, $z_1 z_2 + P(0) = 0$ is a conic along which the Hamiltonian system has toric critical points; this projects to the line $\{(0, b_2) \; : \; b_2 \in \mathbb{R}\}$ under H.

The level sets $H^{-1}(b_1, b_2)$ for $b_2 \neq 0$ are Lagrangian tori, and the level sets $H^{-1}(b_1, 0)$ are Lagrangian tori unless the circle $|z_3|^2 = b_1$ contains a root of P. If this circle contains k roots of P then the fibre $H^{-1}(b_1, 0)$ is a Lagrangian torus with k pinches.

- For example, if $P(z) = z^{n+1} - 1$ then the fibre $H^{-1}(1, 0)$ is the only critical fibre; it has $n + 1$ focus–focus critical points (see Figure 7.4).
- If $0 < a_1 < a_2 < \cdots < a_{n+1}$ are real numbers and $P(z) = (z - a_1)(z - a_2)\cdots(z - a_{n+1})$ then there are $n + 1$ focus–focus fibres which project via H to the points $\{(a_1^2, 0), (a_2^2, 0), \ldots, (a_{n+1}^2, 0)\}$ (see Figure 7.5).

---

[2] Algebraic geometers usually say 'smoothing' to mean the total space of a family which smooths a singularity; some other people say 'smoothing' to mean the smooth fibre of such a family.

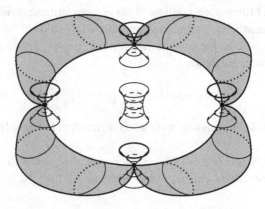

Figure 7.4 The Hamiltonian system from Example 7.6 with $P = z^4 + 1$ ($n = 3$). There is a single fibre with four focus–focus critical points and a smooth conic which consists of toric critical points.

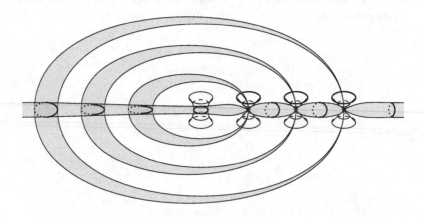

Figure 7.5 The Hamiltonian system from Example 7.6 with $P = (z - a_1)(z - a_2)(z - a_3)$. There is a smooth conic consisting of toric critical points, and three focus–focus fibres which encircle it. We also show, horizontally across the figure, the visible Lagrangian submanifolds described in Remark 7.7.

In any case, the image of the developing map can be analysed as in the Auroux system. We get the diagram shown in Figure 7.6.

Figure 7.6 Almost toric diagram for A_2 Milnor fibre.

The number of base-nodes here is the number of values of b_1 for which $P(z)$ has a zero with $|z|^2 = b_1$. We see that the affine monodromy as we cross the branch cut at position $(t, 0)$ is the product of all the individual affine monodromies for base-nodes with $b_1 < t$, all of which are $\begin{pmatrix} 1 & 0 \\ 1 & 1 \end{pmatrix}$ (just as in the Auroux system). In particular, the 'total monodromy' as we cross the branch cut far to the right is $\begin{pmatrix} 1 & 0 \\ n+1 & 1 \end{pmatrix}$. If we change the branch cut by 180 degrees clockwise then we get the diagram in Figure 7.7 (drawn in the case $n = 2$):

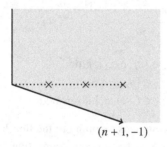

$(n+1, -1)$

Figure 7.7 Another view of Figure 7.6. Here $n = 2$ and there are $n+1$ singularities.

If we apply the integral affine transformation $\begin{pmatrix} 1 & 1 \\ 0 & 1 \end{pmatrix}$ to this diagram, we get Figure 7.8.

Compare this with the moment polygon $\pi(n+1, n)$ from Example 3.21 for the cyclic quotient singularity $\frac{1}{n+1}(1, n)$. Indeed, this is precisely the A_n-singularity

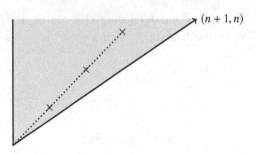

Figure 7.8 A third view of Figure 7.6.

mentioned earlier. As the polynomial $P(z)$ degenerates to z^{n+1}, the base-nodes move in the diagram along the dotted line towards the non-Delzant corner.

Remark 7.7 There are some visible Lagrangians in the Milnor fibre M_P when $P(z) = (z - a_1)(z - a_2) \cdots (z - a_{n+1})$. Namely, consider the antisymplectic involution $(z_1, z_2, z_3) \mapsto (\bar{z}_2, \bar{z}_1, \bar{z}_3)$. The fixed locus consists of points $\{(z_1, \bar{z}_1, z_3) : |z_1|^2 = -P(z_3), z_3 \in \mathbb{R}\}$. The fixed locus of an antisymplectic involution is always a Lagrangian submanifold; in this case, it consists of:

- Lagrangian spheres $z_3 \in [a_k, a_{k+1}]$, k even,
- the Lagrangian plane $z_3 \in (-\infty, a_1]$ if n is even.

The involution $(z_1, z_2, z_3) \mapsto (-\bar{z}_2, \bar{z}_1, \bar{z}_3)$ gives more Lagrangians,

$$\{(z_1, -\bar{z}_1, z_3) : |z_1|^2 = P(z_3), z_3 \in \mathbb{R}\},$$

consisting of

- Lagrangian spheres $z_3 \in [a_k, a_{k+1}]$, k odd,
- the Lagrangian plane $z_3 \in [a_{n+1}, \infty)$,
- the Lagrangian plane $z_3 \in (-\infty, a_1]$ if n is odd.

These are all visible Lagrangians mapping to the line $H_2 = 0$: the Lagrangian spheres project to the compact segments connecting focus–focus fibres; the plane $z_3 \in [a_{n+1}, \infty)$ projects to the segment connecting the rightmost focus–focus fibre to infinity. The plane $z_3 \in (-\infty, a_1]$ has a more singular projection: the disc $z_3 \in [0, a_1]$ projects to the segment connecting the leftmost focus–focus fibre to the toric boundary; the annulus $z_3 \in [-\infty, 0]$ projects to the whole ray $H_2 = 0$ emanating from the toric boundary. In other words, this visible Lagrangian 'folds over itself' at the toric boundary. See Figure 7.9 for the images of these visible Lagrangians under the action map, and Figure 7.5

to see how they look in the total space of the conic fibration. In what follows, the visible disc $z_3 \in [0, 1]$ will be more important; we will denote it by Δ.

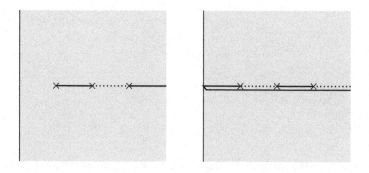

Figure 7.9 The visible Lagrangians described in Remark 7.7 in the case $n = 2$. On the left, we see the components of the fixed locus of $(z_1, z_2, z_3) \mapsto (\bar{z}_2, \bar{z}_1, \bar{z}_3)$: a Lagrangian sphere and a Lagrangian plane. On the right we see the components of the fixed locus of $(z_1, z_2, z_3) \mapsto (-\bar{z}_2, \bar{z}_1, \bar{z}_3)$: a Lagrangian sphere and a Lagrangian plane whose projection 'folds over itself'. In reality, this projection is contained in a single horizontal line; we have separated it for clarity.

7.4 Smoothing Cyclic Quotient T-Singularities

Example 7.8 Let $d \geq 1$ be an integer, p, q be coprime positive integers with $1 \leq q < p$, and $0 < a_1 < \ldots < a_d$ be real numbers. Let P be the polynomial $P(z) = (z^p - a_1)(z^p - a_2) \cdots (z^p - a_d)$. Consider the action of the group μ_p of pth roots of unity on the variety M_P from Example 7.6 given by $\mu \cdot (z_1, z_2, z_3) = (\mu z_1, \mu^{-1} z_2, \mu^q z_3)$, $\mu \in \mu_p$. This action is free and $\pi(\mu \cdot (z_1, z_2, z_3)) = \mu \pi(z_1, z_2, z_3)$.

The Hamiltonian system $\boldsymbol{H}$ on M_P from Example 7.6 has a line of toric critical points along $H_1 = 0$ and d isolated critical fibres with $H_1 = 1, 2, \ldots, d$, each of which has p focus–focus critical points. The μ_p-action preserves the critical fibres: the p focus–focus critical points in each fibre form a μ_p-orbit. The Hamiltonian system $\boldsymbol{H}$ descends to give a system $\boldsymbol{G} \colon B_{d,p,q} \to \mathbb{R}^2$ on the quotient space $B_{d,p,q} := M_P/\mu_p$ with d focus–focus critical points and $\boldsymbol{H}(M_P) = \boldsymbol{G}(B_{p,q})$. However, the action coordinates are different: quotienting by the μ_p-action changes the period lattice (compare with Example 3.21). In fact, a fundamental action domain for $\boldsymbol{G}$ is the polygon $\pi(dp^2, dpq - 1)$, and there are d base-nodes. The branch cut is along a line pointing in the (p, q)-

direction, which is an eigenvector of the affine monodromy; the base-nodes all lie on this branch cut. See Figure 7.10.

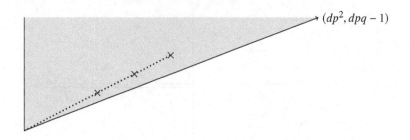

$(dp^2, dpq - 1)$

Figure 7.10 The fundamental action domain for $B_{d,p,q}$, shown in the case $d = 2, p = 2, q = 1$.

Remark 7.9 (Advertising)   The manifold $B_{d,p,q}$ is the Milnor fibre of the cyclic quotient singularity $\frac{1}{dp^2}(1, dpq - 1)$. Cyclic quotient singularities of this form are called *cyclic quotient T-singularities*, and are the most general cyclic quotient surface singularities admitting a $\mathbb{Q}$-Gorenstein smoothing;[3] see [69, Proposition 5.9] or [59, Proposition 3.10]. Perhaps the case that has attracted the most attention is $B_{1,p,q}$, often abbreviated to $B_{p,q}$, because in that case the Milnor fibre has $H_*(B_{1,p,q}; \mathbb{Q}) = H_*(B^4; \mathbb{Q})$, that is, it is a rational homology ball. We will show this shortly. This makes the manifold $B_{1,p,q}$ a useful building block for constructing exotic four-dimensional manifolds with small homology groups, for example using the rational blow-down construction [39].

The symplectic geometry of the manifolds $B_{1,p,q}$ has also been studied. Lekili and Maydanskiy [64] showed that $B_{1,p,q}$ contains no compact exact Lagrangian submanifolds despite having nonzero symplectic cohomology.[4] Karabas [57] showed that the Kontsevich cosheaf conjecture holds for $B_{1,p,1}$, namely that the wrapped Fukaya category of $B_{1,p,1}$ can be calculated using microlocal sheaf theory on the Lagrangian skeleton discussed in Lemma 7.11 later. Evans and Smith [36, 37], building on ideas of Khodorovskiy [58], used obstructions to symplectic embeddings of $B_{1,p,q}$ to obtain restrictions[5] on which cyclic quotient singularities can occur under stable degenerations of complex surfaces.

---

[3] This means that the total space of the smoothing is $\mathbb{Q}$-Gorenstein; this condition picks out a distinguished deformation class of smoothings [59, Theorem 3.9]. Other cyclic quotient singularites can be smoothed, but the total space of the smoothing is not $\mathbb{Q}$-Gorenstein.

[4] The 'standard way' to rule out exact Lagrangian submanifolds is to show that the symplectic cohomology vanishes.

[5] Compare with the algebro-geometric approaches to these problems in the work of Hacking and Prokhorov [51] and the work of Rana and Urzúa [86].

Remark 7.10 As in Remark 7.7, the manifold M_P with $P(z) = (z^p - a_1)(z^p - a_2)\cdots(z^p - a_d)$ contains $p(d-1)$ visible Lagrangian spheres

$$S_{1,1},\ S_{1,2},\ \ldots,\ S_{1,p},\ S_{2,1},\ \ldots,\ S_{d-1,p}.$$

These can be obtained by taking fixed loci as before and applying the μ_p action to the result. When we quotient by μ_p, the spheres descend to give $d-1$ visible Lagrangian spheres $\overline{S}_1, \ldots, \overline{S}_{d-1}$ in $B_{d,p,q}$. We also obtain p visible Lagrangian discs $\Delta_1, \ldots, \Delta_p$ with common boundary along the toric boundary. These discs descend to a visible Lagrangian CW-complex $\overline{\Delta} \subseteq B_{d,p,q}$ which we can think of as a quotient of the unit disc by the equivalence relation which identifies points $z \sim e^{2\pi i q/p} z$ in its boundary. Where $\overline{\Delta}$ meets the toric boundary, it does so along a visible Lagrangian (p,q)-pinwheel core. We call such a Lagrangian CW-complex a *Lagrangian (p,q)-pinwheel*. See Figure 7.11 for the projections of these visible Lagrangians.

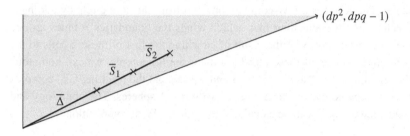

Figure 7.11 The visible spheres and pinwheel in $B_{d,p,q}$, shown in the case $d = 2$, $p = 2$, $q = 1$ (in this case, $\overline{\Delta}$ is an $\mathbb{RP}^2$).

Lemma 7.11 *This union of visible Lagrangians* $\overline{\Delta} \cup \bigcup_{i=1}^{d-1} \overline{S}_i$ *is homotopy equivalent to* $B_{d,p,q}$. *In particular,* $\pi_1(B_{d,p,q}) = \mathbb{Z}/p$, $H_1(B_{d,p,q}; \mathbb{Z}) = \mathbb{Z}/p$ *and* $H_2(B_{d,p,q}; \mathbb{Z}) = \mathbb{Z}^{d-1}$.

Proof The manifold $B_{d,p,q}$ deformation retracts onto the preimage of the line segment ℓ shown in the fundamental action domain in Figure 7.12.

Let ℓ_- and ℓ_+ be the segments of ℓ to the left and right (respectively) of the marked point $\bullet$ in the diagram. The preimage of ℓ_- is a solid torus T_-; our convention will be that the loop $(1,0)$ in ∂T_- bounds a disc in T_-. The preimage of ℓ_+ can be understood as follows. If there were no base-nodes, it would be $T^2 \times [0,1]$. Each base-node means that we pinch the torus above it along a loop in the homology class $(-q, p)$. Up to homotopy equivalence, this is the same as attaching a disc to $T^2 \times [0,1]$ along a loop in this homology class. We

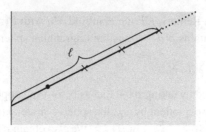

Figure 7.12 A fundamental action domain for $B_{d,p,q}$. The manifold $B_{d,p,q}$ deformation-retracts onto a Lagrangian CW-complex which projects to the line ℓ.

can further homotope these attaching maps so that they attach to loops in the boundary of T_-. Therefore, the preimage of ℓ is homotopy equivalent to a solid torus with d discs attached along its boundary along d parallel copies of the loop $(-q, p)$. Now by a homotopy equivalence we can collapse T_- to its core circle. The result is a CW-complex built up from the core circle by adding d 2-cells using the attaching map which winds the boundaries p times around the core circle. Let $\overline{\Delta}$ be the result of attaching the first of these 2-cells to the core circle. Since all these attaching maps are homotopic, we can homotope the remaining $d - 1$ to a point in $\overline{\Delta}$ and we see that the resulting CW-complex is homotopy equivalent to $\overline{\Delta}$ wedged with $d - 1$ spheres. The homology and fundamental group can be calculated using this CW-decomposition.   □

7.5 Solutions to Inline Exercises

Exercise 7.12   *Given a Hamiltonian system H and a critical point x of H, let $Q(H, x)$ denote the subspace of the space of quadratic forms spanned by the Hessians of the components of H. Check that, after a suitable symplectic change of coordinates, $Q(H, 0) = Q(F, 0)$ where H is the Auroux system and F is the standard focus–focus system from Example 6.1.*

Solution   If we set $z_k = x_k + iy_k$ then $H_1(z_1, z_2) = |z_1 z_2 - c|^2 = 2c(y_1 y_2 - x_1 x_2) + \cdots$, where the dots stand for terms of higher order in the Taylor expansion, and $H_2(z_1, z_2) = \frac{1}{2}(|z_1|^2 - |z_2|^2) = \frac{1}{2}(x_1^2 + y_1^2 - x_2^2 - y_2^2)$. Recall that the standard focus–focus Hamiltonians are

$$F_1 = -p_1 q_1 - p_2 q_2, \quad F_2 = p_2 q_1 - p_1 q_2.$$

If we make the symplectic change of coordinates:

$$p_1 = \frac{1}{\sqrt{2}}(x_2 - y_1) \qquad\qquad p_2 = \frac{1}{\sqrt{2}}(x_1 - y_2)$$

$$q_1 = \frac{1}{\sqrt{2}}(x_1 + y_2) \qquad\qquad q_2 = \frac{1}{\sqrt{2}}(x_2 + y_1)$$

then we get

$$-(p_1 q_1 + p_2 q_2) = y_1 y_2 - x_1 x_2, \quad p_2 q_1 - p_1 q_2 = \frac{1}{2}(x_1^2 - y_1^2 - x_2^2 + y_2^2)$$

so that, to second order, $H_1(z_1, z_2) = 2cF_1(p, q)$ and $H_2(z_1, z_2) = F_2(p, q)$. Therefore, the Hessians of H_1 and H_2 span the same subspace of quadratic forms as the Hessians of F_1 and F_2 after this coordinate change. $\square$

8

Almost Toric Manifolds

We have now seen some examples of Hamiltonian systems with focus–focus critical points; in particular, we have seen what their fundamental action domains look like. We now introduce a definition (*almost toric fibrations*) which covers all of these examples, and use it to develop some general theory for manipulating and interpreting their fundamental action domains (*almost toric base diagrams*).

8.1 Almost Toric Fibration

Definition 8.1 An *almost toric fibration* is a Lagrangian torus fibration $f \colon X \to B$ on a four-dimensional symplectic manifold such that the discriminant locus comprises a collection of zero- and one-dimensional strata such that the smooth structure on B extends over these strata, and f is smooth with respect to this extended smooth structure and has either toric or focus–focus critical points there.

Remark 8.2 We remark that the smooth structure mentioned in the definition plays a somewhat auxiliary role: as in Remark 2.13, the regular locus of the base inherits a, possibly different, smooth structure from its canonical integral affine structure, and this may not extend.

Let $B^{reg} \subseteq B$ be the set of regular values of an almost toric fibration f, let $\tilde{B}^{reg}$ be its universal cover, and let $I \colon \tilde{B}^{reg} \to \mathbb{R}^2$ be the flux map. Let $D \subseteq \tilde{B}^{reg}$ be a fundamental domain for the action of $\pi_1(B^{reg})$. Recall from Remark 6.9 that if $b_1, b_2, \ldots \in B^{reg}$ is a sequence tending to a focus–focus critical point then $\lim_{k \to \infty} I(b_k)$ is a well-defined point in $\mathbb{R}^2$ called the *base-node* of that critical point. There is also an affine monodromy associated to loops in B^{reg} that go around the base-node. Lemma 6.13 tells us that this monodromy is completely determined by specifying, at each base-node, a primitive integral eigenvector (p, q) with eigenvalue 1 for the monodromy.

Definition 8.3 The *almost toric base diagram* associated to these choices is the fundamental action domain $I(D) \subseteq \mathbb{R}^2$ decorated with the positions of the base-nodes and the eigenvector at each base node.

Remark 8.4 We will usually (though not always) choose our fundamental domain D by making branch cuts connecting the base nodes to the boundary along eigenlines.

Although the almost toric base diagram does not determine $f: X \to B$ up to *fibred* symplectomorphism, the next theorem guarantees that it does determine X up to symplectomorphism.

Theorem 8.5 (Symington [106, Corollary 5.4]) *Suppose that $f: X \to B$ and $g: Y \to B$ are almost toric fibrations whose almost toric base diagrams are the same. If B is a punctured two-dimensional surface then X and Y are symplectomorphic.*

Proof Let $N \subseteq B$ be the set of base-nodes. By Theorem 6.17, there is a neighbourhood U of N together with a symplectomorphism $\Phi: f^{-1}(U) \to g^{-1}(U)$. Although this symplectomorphism is not fibred, it is fibred near the boundary of U. Choose Lagrangian sections of f and g over U (for example, we can use the section $\sigma_1(b) = (-\bar{b}, 1)$ in each focus–focus chart,; see the proof of Theorem 6.7). If we can find global Lagrangian sections over $B \setminus N$ which match with the chosen Lagrangian section over U along $U \setminus N$ then Theorem 2.26 gives us a fibred symplectomorphism $f^{-1}(B \setminus U) \to g^{-1}(B \setminus U)$ extending the symplectomorphism $\Phi: f^{-1}(U) \to g^{-1}(U)$. Since $B \setminus N$ has the homotopy type of a punctured surface and N is a strict subset of the punctures (by assumption), the relative cohomology group $H^2(B \setminus N, U \setminus N)$ vanishes. Therefore, Corollary 2.33 tells us we can extend the Lagrangian section as required. (This was the strategy we alluded to in Remark 2.32.) □

8.2 Operation I: Nodal Trade

We now introduce some tools for manipulating and constructing almost toric base diagrams which will give us a wealth of examples. The first of these is Symington's *nodal trade*. It allows us to 'trade' a Delzant corner for a base-node.

Recall from Figure 7.3 that there is an almost toric structure on $\mathbb{C}^2$ which admits a fundamental action domain as drawn on the left in Figure 8.1(a). The shaded region is integral affine equivalent to the shaded region in Figure 8.1(b), which is a subset of the moment polygon for the standard torus action on $\mathbb{C}^2$. This means that the preimages of these two regions are fibred-symplectomorphic.

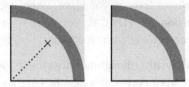

Figure 8.1 (a) A fundamental action domain for the Auroux system on $\mathbb{C}^2$. (b) The moment image for the standard torus action on $\mathbb{C}^2$. The shaded subsets in both diagrams are integral affine equivalent to one another.

In particular, whenever we see a Delzant corner, we can excise it and glue in a copy of the Auroux system, using this fibred symplectomorphism to make identifications. Since the identifications are fibred, this operation yields a new Lagrangian torus fibration on the same manifold.[1] In fact, there are many different operations, one for each Vũ Ngọc model, but the results are all (non-fibred) symplectomorphic to one another by Theorem 6.17. We call an operation like this a *nodal trade*.

Remark 8.6   The toric boundary near a Delzant corner comprises two symplectic discs meeting transversally at the vertex. When you perform a nodal trade, the toric boundary becomes a symplectic annulus which is a smoothing of this pair of discs. For example, in the Auroux system this is the smoothing from $z_1 z_2 = 0$ to $z_1 z_2 = c$.

Example 8.7   Here are some Lagrangian torus fibrations on $\mathbb{CP}^2$:

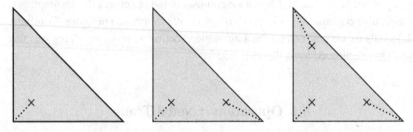

The nodal trade in the lower left corner should look familiar; we call this a *standard Delzant corner*. To find the eigendirection for *any* Delzant corner

---

[1] To see that the manifold does not change, observe that we are excising a symplectic ball and gluing in another symplectic ball with the same boundary (a contact 3-sphere). The contactomorphism group of the 3-sphere is connected [29], so there is a unique way to glue up to isotopy.

p, if A is the unique integral affine transformation which maps the standard Delzant corner to p then the eigendirection at p is $(1, 1)A$. For example, the top left corner is the image of the standard Delzant corner under $\begin{pmatrix} 0 & -1 \\ 1 & -1 \end{pmatrix}$, so the eigendirection is $(1, -2)$, as shown.

As noted in Remark 7.4, although the toric boundary looks like three line segments, every time it crosses a branch cut you have to apply the affine monodromy to its tangent vector, so the apparent breaks in the line when it crosses a branch cut are just an illusion: it is really an uninterrupted straight line in the affine structure. In the three preceding examples, the toric boundary comprises

- a conic and a line (two spheres intersecting transversely at two points, one having twice the symplectic area of the other),
- a nodal cubic curve (pinched torus having symplectic area three),
- a smooth cubic curve (torus having symplectic area three).

This should make sense: the toric boundary for the usual toric picture of $\mathbb{CP}^2$ comprises three lines and the preceding configurations are obtained by smoothing one or more intersections between these lines. Although I have used the terminology 'line', 'conic', and 'cubic' from algebraic geometry, it is not clear for these new integrable Hamiltonian systems whether the toric boundary is actually a subvariety for the standard complex structure. It is, at least, a symplectic submanifold (immersed, where there are double points), and it is known that low-degree symplectic surfaces in $\mathbb{CP}^2$ are isotopic amongst symplectic surfaces to subvarieties (see Gromov [44] for smooth surfaces of degrees 1 and 2, Sikorav [102] for smooth surfaces of degree 3, Shevchishin [99] for nodal surfaces of genus at most 4, and Siebert and Tian [100, 101] for smooth surfaces in degrees less than or equal to 17), hence the abuse of terminology.

The diagrams we have drawn rely on a specific choice of fundamental domain D in the universal cover of the regular locus of the almost toric fibration. If we simply plot the image of the developing map (flux map) on the whole of $\tilde{B}^{reg}$, we get some very beautiful pictures which are symmetric under the action of $\pi_1(B^{reg})$ via affine monodromy. In Figures 8.2–8.4 you can see what this looks like for the three examples in Example 8.7. These pictures are closely related to the idea of *mutation* we will meet in Section 8.4, and the *Vianna triangles* that we will see later (Theorem 8.21 and Appendix I).

Figure 8.2 The image of the developing map for an almost toric structure on $\mathbb{CP}^2$ obtained from the standard moment triangle by a single nodal trade in the lower left corner.

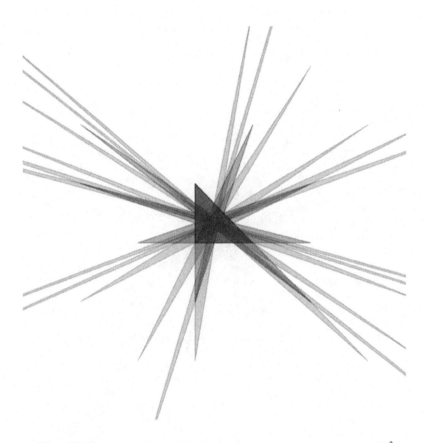

Figure 8.3 The image of the developing map for an almost toric structure on $\mathbb{CP}^2$ obtained from the standard moment triangle by a nodal trade in the lower left corner and a nodal trade in the lower right corner.

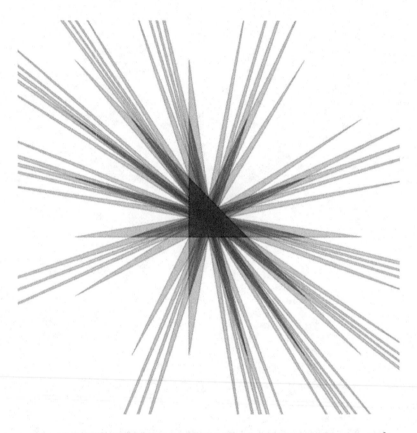

Figure 8.4 The image of the developing map for an almost toric structure on $\mathbb{CP}^2$ obtained from the standard moment triangle by nodal trades in all three corners.

Remark 8.8   A closely related operation takes an almost toric base diagram with a (non-Delzant) corner modelled on the polygon $\pi(dp^2, dpq - 1)$ and replaces it with the diagram in Figure 7.10. We will call this a *generalised nodal trade*. One can think of this as smoothing the cyclic quotient T-singularity in the original almost toric orbifold.

Example 8.9   Take the cubic surface in $\mathbb{CP}^3$ given in homogeneous coordinates $[z_1 : z_2 : z_3 : z_4]$ by $z_1 z_2 z_3 = z_4^3$. This has three A_2-singularities at $[1 : 0 : 0 : 0]$, $[0 : 1 : 0 : 0]$, and $[0 : 0 : 1 : 0]$. It is toric, with the torus action given by $(z_1, z_2, z_3, z_4) \mapsto (e^{3i\theta_1} z_1, e^{3i\theta_2} z_2, z_3, e^{i(\theta_1 + \theta_2)} z_4)$. The Hamiltonians $H_1 = \frac{3|z_1|^2 + |z_4|^2}{|z|^2}$ and $H_2 = \frac{3|z_2|^2 + |z_4|^2}{|z|^2}$ (with $|z|^2 = \sum_{i=1}^{4} |z_i|^2$) generate this action and their image is the triangle $\{(b_1, b_2) \in \mathbb{R}^2 : b_1 \geq 0, b_2 \geq 0, b_1 + b_2 \leq 3\}$. However, the period lattice is not standard; for example, the element $\theta_1 = 2\pi/3$ $\theta_2 = 4\pi/3$ acts trivially. If we use the Hamiltonians H_2 and $(H_1 + 2H_2)/3$, whose period lattice is standard, then the moment polygon becomes the triangle with vertices at $(0, 0)$, $(0, 1)$ and $(3, 2)$. This has three corners, each modelled on the polygon $\pi(3, 2)$, corresponding to the cyclic quotient singularity $\frac{1}{3}(1, 2)$ (also known as A_2). Performing three generalised nodal trades gives the almost toric base diagram for the smooth cubic surface shown in Figure 8.5.

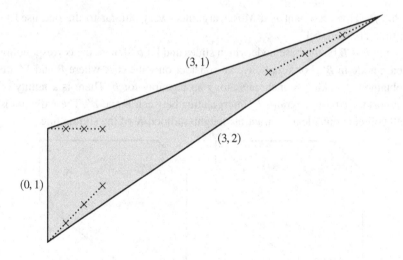

Figure 8.5 An almost toric base diagram for a cubic surface; the edges are labelled with primitive integer vectors pointing along them.

8.3 Operation II: Nodal Slide

Note that there is a free parameter $c > 0$ in the Auroux system. As this parameter varies, we obtain a family of Lagrangian torus fibrations in which the focus–focus critical point moves in the direction of the eigenvector for its affine monodromy (see Figure 8.6). Such a family of fibrations is called a *nodal slide*.

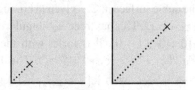

Figure 8.6 A nodal slide.

Theorem 8.10 (Symington's theorem on nodal slides [106, Theorem 6.5]) *Suppose that X and X' are almost toric manifolds whose almost toric base diagrams B and B' are related by a nodal slide. Suppose that the base diagrams have the homotopy type of a punctured two-dimensional surface. Then X and X' are symplectomorphic.*

Proof This argument is a Moser argument very similar to the one used to prove Theorem 6.17.

Let $b \in B$ be the base node which slides and let b' denote the corresponding base node in B'. For simplicity, we consider only the case where B and B' are obtained by taking branch cuts along an eigenray for b. There is a family of almost toric base diagrams B_t interpolating between B and B'. These diagrams all coincide outside a contractible neighbourhood K of the sliding line.

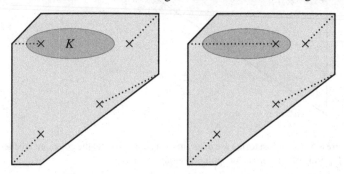

This gives a family of almost toric manifolds X_t together with almost toric fibrations $f_t \colon X_t \to B_t$. Since the diagrams B_0 and B (respectively B_1 and B') are identical, X_0 and X (respectively X_1 and X') are symplectomorphic by Theorem 8.5. (Here is the first place where we use the assumption on the topology of the base.)

It suffices to show that X_0 and X_1 are symplectomorphic. We use the fact that the almost toric bases coincide outside K to identify the subsets $U_t = f_t^{-1}(B_t \setminus K)$, extend this to a family of diffeomorphisms $\varphi_t \colon X_0 \to X_t$, and show that the symplectic forms $\omega_t = \varphi_t^* \omega_{X_t}$ on X_0 satisfy $d[\omega_t]/dt = 0$. The result will then follow from Moser's argument (Appendix E). The subset on which the symplectic forms differ is $f_t^{-1}(K)$, which is just a neighbourhood of the nodal fibre. As we saw in the proof of Theorem 6.17, it suffices to show that $\int_{\sigma_t} d\omega_t/dt = 0$ for some family of submanifolds σ_t such that $\varphi_t(\sigma_t)$ intersects the nodal fibre of f_t once transversely and $\sigma_t = \sigma_0$ on $\varphi_t^{-1}(U_t)$. As in that proof, we can use a chosen family of Lagrangian sections. (Here again we use the topology of the base to guarantee the existence of these sections.) $\quad\square$

As a result, nodal sliding does not change the symplectic manifold, but it certainly changes the Lagrangian fibration. Here is an example.

Example 8.11 Start with the almost toric fibration on $\mathbb{CP}^2$ from Example 8.7 with three base nodes. Pick the bottom-right node (labelled B in Figure 8.7), and slide it towards the opposite edge, beyond the barycentre of the triangle. We get two almost toric fibrations on $\mathbb{CP}^2$ which we can distinguish as follows. Consider the Lagrangian vanishing thimble emanating from the base-node labelled A: this is a visible Lagrangian disc with centre at the focus–focus critical point, and its projection is a ray in the base diagram pointing in the eigendirection of this node, which is $(1, 1)$.

Before the nodal slide, this ray hits the slanted edge. It is an exercise (Exercise 8.34) to show that the visible Lagrangian meets this edge along a $(2, 1)$-pinwheel core, that is, a Möbius strip.[2] After the nodal slide, this ray crosses the sliding branch cut. When it emerges its direction has changed by the affine monodromy $\begin{pmatrix} -1 & 1 \\ -4 & 3 \end{pmatrix}$ so that the ray now points in the $(-5, 4)$-direction. Exercise 8.34 also asks you to show that this ray now intersects the slanted edge of the base diagram in a $(5, 4)$-pinwheel core. This difference in topology distinguishes the torus fibrations.

Remark 8.12 Visible Lagrangians obtained by capping a (p, q)-pinwheel core with a disc are quite common in this context, and we call them *Lagrangian (p, q)-pinwheels*.

---

[2] Indeed, this visible Lagrangian is just $\mathbb{RP}^2 \subseteq \mathbb{CP}^2$.

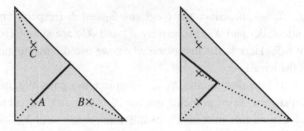

Figure 8.7 Left: Before the nodal slide, the vanishing thimble of base-node A is part of a visible Lagrangian $\mathbb{RP}^2$. Right: After the nodal slide, it is part of a visible Lagrangian $(5, 4)$-pinwheel.

The curious reader may wonder what happens if we try to slide a node in a direction which is not its eigenline. If we do so, we obtain a family of almost toric base diagrams, and so a family of symplectic manifolds. However, the cohomology class of the symplectic form can vary, so they are not all symplectomorphic. Here is an example.

Example 8.13 Take the toric diagram for $O(-1)$ giving the compact edge affine length 2 and make two nodal trades.

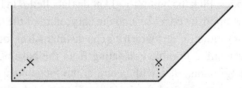

Now we attempt to slide the right-hand node to the left. As it moves, more and more of what used to be the compact edge passes through the branch cut and ends up as part of the slanted noncompact edge on the right.

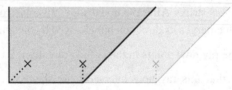

If we 'undo' the nodal trades, we see that the compact edge has shrunk. In other words, the symplectic area of the zero-section has decreased, and the cohomology class of ω has changed.

Indeed, one can use this to 'flip' the sign of the symplectic area of curves. This is closely related to the theory of flips in algebraic geometry; see [38] for a full discussion with examples.

8.4 Operation III: Mutation

In Example 8.11, we needed to keep track of a visible Lagrangian which crossed a branch cut. This can get very tricky if there are several branch cuts. Sometimes, it is more convenient to change the choice of fundamental domain in $\tilde{B}^{reg}$. We will do this by rotating the branch cut as we did in Section 7.2 and Figure 7.2.

In fact, in most examples, we will start with a branch cut which emanates from a base-node in the direction v of an eigenvector for the affine monodromy, and rotate by 180 degrees anticlockwise (or clockwise) to get a branch cut in the $-v$-direction.

The fundamental action domain (almost toric base diagram) transforms in the following way. The eigenline bisects the diagram; call the two pieces D_1 and D_2 (we adopt the convention that D_1 lies clockwise of the branch cut and D_2 lies anticlockwise). Let M be the affine monodromy around the base-node as we cross the branch cut in the anticlockwise direction. When the branch cut rotates anticlockwise 180 degrees, we replace D_2 with $(D_2)M^{-1}$ to get the new almost toric base diagram $D_1 \cup (D_2)M^{-1}$. When the branch cut rotates clockwise 180 degrees, we get $(D_1)M \cup D_2$ instead.

Remark 8.14   Note that changing branch cut has no effect on the symplectic manifold nor on the Lagrangian torus fibration. All that changes is the picture: the picture is the image of a fundamental action domain under the developing map, and the change of branch cut amounts to a different choice of fundamental action domain.

Example 8.15   Take the following almost toric base diagram and let x be the base-node marked B. The anticlockwise affine monodromy is $M = \begin{pmatrix} -1 & 1 \\ -4 & 3 \end{pmatrix}$ and the branch cut points in the $(2, -1)$-direction. We have indicated the coordinates of the corners of the triangle. The affine lengths of the three edges are 3; this choice corresponds to the pullback of the Fubini–Study form along the anticanonical embedding of $\mathbb{CP}^2$ in $\mathbb{CP}^9$.

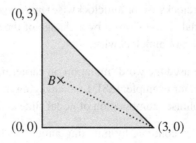

If we rotate the associated branch cut 180 degrees anticlockwise, then the result is:

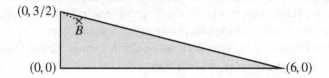

We superimpose the two pictures for easier comparison.

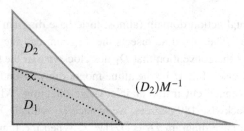

After performing this change of branch cut, the visible Lagrangian $(5,4)$-pinwheel from Example 8.11 is easier to see, as its projection does not cross any branch cuts (see Figure 8.8).

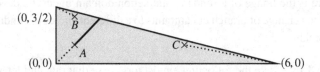

Figure 8.8 The visible $(5,4)$-pinwheel is easier to see after applying a mutation to Figure 8.7. Note that the direction of the branch cut at C is obtained from the original cut by applying the clockwise B-monodromy. These cuts are hard to see, as they are almost parallel to the edges. The vectors indicate the coordinates of the vertices.

Definition 8.16 A (clockwise or anticlockwise) *mutation* at a base-node is a combination of a nodal slide followed by a change of the same branch cut by 180 degrees (clockwise or anticlockwise).

Remark 8.17 I have used the word 'mutation' in earlier papers to mean just a change of branch cut, for example, [38]. I am changing my preference here to avoid overusing the phrase 'combination of nodal slide and mutation'.

Example 8.18 Let us continue by mutating anticlockwise at the base-node labelled C in Figure 8.8. We now find a visible Lagrangian $(13, 2)$-pinwheel living over the eigenray emanating from the base-node C.

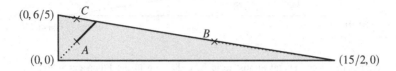

Figure 8.9 After a further mutation, we see a visible $(13, 2)$-pinwheel. By now, the B and C branch cuts are indistinguishably close to the edges.

Remark 8.19 One useful trick for figuring out the directions of the branch cuts is the following. In the original $\mathbb{CP}^2$ triangle, the eigenlines all intersect at the barycentre of the triangle $(1, 1)$. This remains true after mutation. So the directions of the branch cuts can be found by taking the vectors from $(1, 1)$ to the corners. For example, the B branch cut in Figure 8.9 points in the $(13, -2)$-direction. This trick works when all the eigenlines intersect at a common point. This is a cohomological condition; see Section 8.6.

One could continue in this fashion, mutating at B, then C, then B, and so on. This would give a sequence of Lagrangian (p_k, q_k)-pinwheels in $\mathbb{CP}^2$ where p_k runs over the odd-indexed Fibonacci numbers.[3] The general almost toric base diagram in this sequence is shown in Figure 8.10; to declutter the diagram we have 'undone' the nodal trade at the vertex which remains unmutated. The shaded region in this figure is a symplectically embedded copy of the solid ellipsoid

$$\left\{ (z_1, z_2) \in \mathbb{C}^2 \; : \; \frac{1}{2} \left(\frac{F_{2n-1}}{3F_{2n+1}} |z_1|^2 + \frac{F_{2n+1}}{3F_{2n-1}} |z_2|^2 \right) \leq \lambda \right\}$$

for $\lambda < 1$. We can get an embedding with λ arbitrarily close to 1 by nodally sliding the two base-nodes very close to their vertices and pushing the slanted edge of the shaded region towards the toric boundary. In other words, we can fill up an arbitrarily large fraction of the volume of $\mathbb{CP}^2$ by a symplectically embedded ellipsoid with this Fibonacci ratio of radii. This is related to the famous *Fibonacci staircase* pattern for full ellipsoid embeddings in $\mathbb{CP}^2$ observed by McDuff and Schlenk [78]. For papers which discuss this and other examples from an almost toric point of view, see [8, 11, 22, 70].

In fact, one could also allow sequences of mutations involving base-node A, to obtain an infinite trivalent tree of almost toric base diagrams all representing

[3] Our convention is that $F_1 = 1$, $F_3 = 2$, $F_5 = 5$, and so on.

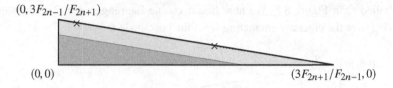

$(0, 3F_{2n-1}/F_{2n+1})$

$(0,0)$ $(3F_{2n+1}/F_{2n-1}, 0)$

Figure 8.10 The general almost toric base diagram in the Fibonacci staircase. The shaded region is a solid ellipsoid which can be rescaled to fill an arbitrarily large fraction of the volume of $\mathbb{CP}^2$ provided we slide the base-nodes closer to the corners.

almost toric fibrations on $\mathbb{CP}^2$. This tree is closely related to another famous infinite trivalent tree arising from the theory of Diophantine approximation: the *Markov tree*. For the (p, q)-pinwheels which appear, p is a Markov number.[4]

Definition 8.20 A *Markov triple* is a positive integer solution p_1, p_2, p_3 to the *Markov equation*

$$p_1^2 + p_2^2 + p_3^2 = 3p_1p_2p_3.$$

A Markov number is a number which appears in a Markov triple.

Theorem 8.21 (Vianna [116]) *For every Markov triple p_1, p_2, p_3, there is an almost toric diagram $D(p_1, p_2, p_3)$ (a Vianna triangle) with the following properties.*

- *The diagram $D(1, 1, 1)$ is obtained from the standard toric diagram of $\mathbb{CP}^2$ by performing three nodal trades.*
- *The diagram $D(p_1, p_2, p_3)$ is a triangle with three base-nodes n_1, n_2, n_3, obtained by iterated mutation on $D(1, 1, 1)$ (in particular, the associated almost toric manifold is $\mathbb{CP}^2$).*
- *For $k = 1, 2, 3$, there is an integer q_k and a Lagrangian pinwheel of type (p_k, q_k) living over the branch cut which connects n_k to a corner P_k.*
- *The affine length of the edge opposite the corner P_k is $3p_k/(p_{k+1}p_{k+2})$ where indices are taken modulo 3.*

Proof We prove this in Appendix I, where we also remind the reader of the basic properties of Markov triples.   □

Remark 8.22   The diagrams in Figures 8.8 and 8.9 are $D(1, 1, 2)$ and $D(1, 2, 5)$.

Remark 8.23   Superimposing all these almost toric base diagrams yields Figure

---

[4] The odd-indexed Fibonacci numbers are a subset of the Markov numbers.

8.4, that is, the image of the developing map of the integral affine structure on the universal cover of the complement of the base-nodes.

Remark 8.24 For each diagram $D(p_1, p_2, p_3)$, let $T(p_1, p_2, p_3)$ denote[5] the Lagrangian torus fibre over the barycentre. Before the work of Vianna, the tori $T(1, 1, 1)$ (the *Clifford torus*) and $T(1, 1, 2)$ (the *Chekanov torus*) had been constructed and Chekanov [16] had shown that they were not Hamiltonian isotopic. Vianna's truly remarkable contribution, besides constructing $T(p_1, p_2, p_3)$, was to show that if p_1, p_2, p_3 and p_1', p_2', p_3' are distinct as unordered Markov triples then $T(p_1, p_2, p_3)$ and $T(p_1', p_2', p_3')$ are not Hamiltonian isotopic.[6]

Remark 8.25 Vianna [117] has also studied mutations of other triangular almost toric base diagrams and used this to construct exotic Lagrangian tori in other symplectic 4-manifolds.

Remark 8.26 As far as I know, it is a completely open question to characterise which quadrilaterals arise as mutations of a square or a rectangle.

8.5 More Examples

Example 8.27 In Example 8.15, we constructed the almost toric base diagram for $\mathbb{CP}^2$ shown in Figure 8.11 by performing one nodal trade and a mutation (we have nodally slid the base node to make it clearer). The visible Lagrangian pinwheel over the branch cut is $\mathbb{RP}^2$ and the symplectic sphere living over the opposite edge is isotopic to a conic curve (its symplectic area is twice that of a complex line in $\mathbb{CP}^2$, so it inhabits the homology class of a conic, and symplectic curves in this class are known to be isotopic [44]). If we excise the conic, what is left is a subset of the almost toric base diagram for $B_{1,2,1}$ from Example 7.8. This is actually the cotangent bundle of $T^*\mathbb{RP}^2$. Thus, we see that the complement of a conic in $\mathbb{CP}^2$ is symplectomorphic to a neighbourhood of $\mathbb{RP}^2$ in its cotangent bundle.

Remark 8.28 (Exercise 8.35) Using the same idea, we find that if $X = S^2 \times S^2$ with its equal-area symplectic form and $C \subseteq X$ is a symplectic sphere isotopic to the diagonal then $X \setminus C$ is a neighbourhood of a Lagrangian sphere in its cotangent bundle. Of course, these results can be proved directly and explicitly, but almost toric diagrams give us a way to see them and generalise them to less obvious examples.

---

[5] Vianna [116] uses the notation $T(p_1^2, p_2^2, p_3^2)$.

[6] He had an earlier paper [115] which distinguished $T(1, 2, 5)$ from $T(1, 1, 1)$ and $T(1, 2, 5)$, which was already a major breakthrough. At the time it appeared, I was convinced there should be only two Hamiltonian isotopy classes of monotone Lagrangian tori in $\mathbb{CP}^2$.

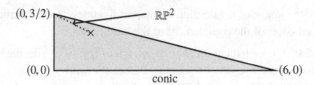

Figure 8.11 Almost toric diagram of $\mathbb{CP}^2$ with a visible Lagrangian $\mathbb{RP}^2$.

Example 8.29 The complement of a cubic curve in $\mathbb{CP}^2$ can be given the almost toric diagram shown in Figure 8.12(a). This is a Weinstein domain which retracts onto a visible Lagrangian cell complex, coloured in the figure: we take the fibre over the barycentre, together with the three Lagrangian vanishing thimbles coming from Lemma 6.15 living over the three lines connecting the base-nodes to the barycentre. The attaching maps for these discs are loops in the barycentric torus living in the homology classes $(1, 2)$, $(1, 2)$ and $(1, -1)$. In Figure 8.12(b), we draw these three loops in a square-picture of the barycentric torus. This Lagrangian cell complex is called the *Lagrangian skeleton* of the complement of the cubic. The paper [1] explores in more detail how to read Weinstein handlebody decompositions off from almost toric base diagrams, and [97] explains how a torus with Lagrangian discs attached is all you need to recover the full complexity of Vianna's tori.

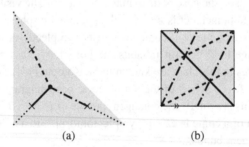

(a)   (b)

Figure 8.12 (a) An almost toric fibration on the complement of a cubic in $\mathbb{CP}^2$.
(b) The boundaries of the three visible discs on the barycentric torus (thought of as a square with its sides identified in opposite pairs).

8.6 Cohomology Class of the Symplectic Form

We have seen that the symplectic area of a curve in the toric boundary of a toric manifold is given by 2π times the affine length of the edge to which it projects. Since the cohomology class of the symplectic form is determined by its integrals over curves, this is enough to determine the cohomology class of the symplectic form on a toric manifold. We now come up with a prescription for finding the cohomology class of the symplectic form on an almost toric manifold. We focus on a restricted class of base diagrams where the almost toric base is $\mathbb{R}^2$ with some base-nodes; one can easily extend our analysis, for example, to the case where there are toric critical points.

Remark 8.30   The case where the base is a closed 2-manifold like the sphere is trickier because it is impossible to 'read off' the symplectic areas of sections. For example, one can change the symplectic area of a section without changing the almost toric base diagram by pulling back a non-exact 2-form from the base of the fibration and adding it to the symplectic form.

Here is the class of almost toric base diagrams on which we will focus. Suppose $f\colon X \to B$ is an almost toric fibration where the almost toric base diagram is $\mathbb{R}^2$ with a collection of base-nodes $n_1, \ldots, n_k$ and branch cuts. Since we are only interested in the integral affine structure, we can choose which point we consider to be the origin 0. We can ensure that $0 \notin \{n_1, \ldots, n_k\}$ and that the straight line segments s_i connecting 0 to n_i intersect only at 0. We write the positions of the base-nodes as (x_i, y_i) relative to this origin. Suppose that the branch cut at n_i points in the (a_i, b_i)-direction, and suppose that this is a primitive integer eigenvector of the affine monodromy at n_i. We will suppose for simplicity that none of the line segments s_i cross the branch cuts.

Theorem 8.31   *The second cohomology (with $\mathbb{R}$-coefficients) of X is the cokernel of the map $\partial\colon \mathbb{R}^2 \to \mathbb{R}^{k+1}$ defined by*

$$\partial(v, w) = (0, b_1 v - a_1 w, \ldots, b_k v - a_k w).$$

If we write $[z_0, z_1, \ldots, z_k]$ for the equivalence class of $(z_0, z_1, \ldots, z_k)$ in this cokernel, then the symplectic form lives in the cohomology class $[0, 2\pi(b_1 x_1 - a_1 y_1), \ldots, 2\pi(b_k x_k - a_k y_k)]$.

Proof   The space X deformation retracts onto the following CW complex W. Let F be the torus fibre over the origin and let s_i be the line segment connecting 0 to n_i. Inside $f^{-1}(s_i)$ there is a disc Δ_i with boundary on F and with centre at the focus–focus critical point over n_i. The boundary of this disc is a loop in F which inhabits the class $(-b_i, a_i) \in H_1(F; \mathbb{Z})$. Let $W = F \cup \bigcup_{i=1}^{k} \Delta_i$.

The fact that X retracts onto W can be proved using Morse theory (though we only sketch it here). The Morse–Bott function $|f|^2$ has a global minimum along F and index 2 critical points at the focus–focus points. The discs Δ_i are

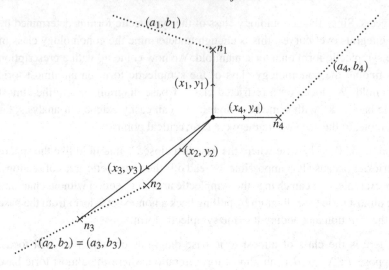

Figure 8.13 A typical almost toric base diagram of the kind we consider.

the downward manifolds emanating from these critical points, which gives a handle-decomposition of X with W as the union of cores of handles.

We pick a CW structure on F with one 0-cell, two 1-cells, and one 2-cell Δ_0. To get a CW structure on W we need to homotope the attaching maps for the 2-cells Δ_i so that they attach to the 1-skeleton of F (which we can do using cellular approximation). Now the cellular chain complex computing $H^2(X; \mathbb{R})$ is

$$C_2(X; \mathbb{R}) = \mathbb{R}^{k+1} \xrightarrow{\partial} C_1(X; \mathbb{R}) = \mathbb{R}^2 \xrightarrow{\partial} C_0(X; \mathbb{R}) = \mathbb{R}.$$

The map $\partial : C_2 \to C_1$ is given by looking at the boundaries of the 2-cells: $\partial\Delta_0 = 0$ and $\partial\Delta_i = (-b_i, a_i)$. We can think of this as a 2-by-$(k+1)$ matrix

$$\begin{pmatrix} 0 & -b_1 & \cdots & -b_k \\ 0 & a_1 & \cdots & a_k \end{pmatrix}.$$

When we take the dual complex (to compute cohomology), the differential $\partial : C^1 \to C^2$ is given by the transpose of this matrix. The cohomology is the cokernel of this map (by definition).

We can compute the integrals of the symplectic form over the 2-cells Δ_i, which then tells us its cohomology class. In action-angle coordinates away from the focus–focus point, Δ_i is the cylinder $[0, 1) \times S^1 \ni (s, t) \mapsto (x_i s, y_i s, -b_i t, a_i t)$, which has symplectic area

$$\int_0^1 \int_{S^1} (a_i y_i - b_i x_i) ds \wedge dt = 2\pi(a_i y_i - b_i x_i)$$

as required. □

Corollary 8.32 *The symplectic form on X is exact if there is a common point of intersection of all the eigenlines from the base-nodes.*

Proof If we were to use this common point of intersection as our origin, it would give all the discs Δ_i area zero because each vector (x_i, y_i) would be proportional to (a_i, b_i). □

Remark 8.33 For readers who are interested in disc classes with boundary on Lagrangian tori, this also tells us that if there is a common intersection point of the eigenlines then the Lagrangian torus fibre over this point is exact. More generally, if there is a toric boundary divisor, this Lagrangian torus will be *monotone* provided the ambient manifold is monotone. In particular, Vianna's tori are all monotone.

8.7 Solutions to Inline Exercises

Exercise 8.34 *Show that the visible Lagrangians in the following pictures have respectively $(2, 1)$- and $(5, 4)$-pinwheel cores where they meet the edge. In the first diagram, the direction of the line in the base is $(1, 1)$. In the second diagram, the line points in the $(-5, 4)$-direction after crossing the branch cut.*

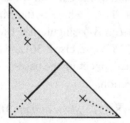 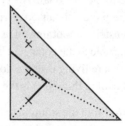

Solution In each case, let ℓ be the primitive integral vector pointing along the line of the visible Lagrangian where it intersects the edge, and let e be the primitive integral vector pointing along the edge. Make an integral affine change of coordinates M so that $eM = (1, 0)$. By comparison with Example 5.14, we see that if $\ell M = \pm(q, p)$ then we have a visible Lagrangian (p, q)-core (here q is determined only up to adding a multiple of p).

In the first case, we have $\ell = (-1, -1)$ and $e = (-1, 1)$, so we use $M = \begin{pmatrix} -1 & -1 \\ 0 & -1 \end{pmatrix}$ which gives $\ell M = (1, 2)$.

In the second case, we have $\ell = (5, -4)$ and $e = (0, -1)$, so we use $M = \begin{pmatrix} 0 & 1 \\ -1 & 0 \end{pmatrix}$ which gives $\ell M = (4, 5)$. □

Exercise 8.35 (Remark 8.28) *Using the same idea as in Example 8.27, show that if $X = S^2 \times S^2$ with its equal-area symplectic form and $C \subseteq X$ is a symplectic sphere isotopic to the diagonal then $X \setminus C$ is a neighbourhood of a Lagrangian sphere in its cotangent bundle.*

Solution   Figure 8.14 shows what you get when you start with the standard square moment map picture of $S^2 \times S^2$, perform nodal trades at two opposite corners, and then mutate one of them.

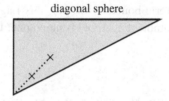

Figure 8.14 An almost toric picture of $S^2 \times S^2$. The complement of the diagonal sphere is a neighbourhood of the visible Lagrangian sphere living over the dotted line connecting the two base-nodes.

After the nodal trades, the toric boundary consists of two symplectic spheres, each isotopic to the diagonal. The complement of the one marked as 'diagonal sphere' in the picture is an open subset of $B_{2,1,1}$, which is the Milnor fibre of a $\mathbb{Q}$-Gorenstein smoothing of a $\frac{1}{2}(1, 1)$ (or A_1) singularity. But $B_{2,1,1}$ is symplectomorphic to the cotangent bundle T^*S^2 (Exercise 8.36) and the visible Lagrangian sphere living over the dotted line connecting the base-nodes (made up of two vanishing thimbles) is the zero-section. □

Exercise 8.36 (From the solution to Exercise 8.35) *Show that the Milnor fibre of the A_1 singularity is symplectomorphic to the cotangent bundle of S^2.*

Solution   The Milnor fibre is an affine quadric $z_1 z_2 + z_3^2 = 1$. We will write an explicit symplectomorphism between the affine quadric and T^*S^2. First, we make a change of coordinates $z_1 = \xi_1 + i\xi_2$, $z_2 = \xi_1 - i\xi_2$, $z_3 = \xi_3$ to get the quadric in the form

$$\xi_1^2 + \xi_2^2 + \xi_3^2 = 1.$$

If we think of S^2 as sitting inside $\mathbb{R}^3$ as the unit sphere $q_1^2 + q_2^2 + q_3^2 = 1$ then its cotangent bundle sits inside $\mathbb{R}^3 \times \mathbb{R}^3$ and consists of pairs $(p, q) \in \mathbb{R}^3 \times S^2$ such that $\sum_{k=1}^{3} p_k q_k = 0$. Here, we give $\mathbb{R}^3 \times \mathbb{R}^3$ the symplectic structure $\sum_{k=1}^{3} dp_k \wedge dq_k$. Now the map sending $\xi = x + iy$ to $(p, q) = (-|x|y, x/|x|)$ is a symplectomorphism. To see this, write $r = |x|$ and pull back the 2-form $\sum dp_k \wedge dq_k$:

$$dp_k \wedge dq_k = -(y_k \, dr + r \, dy_k) \wedge \left(\frac{dx_k}{r} - \frac{x_k \, dr}{r^2} \right)$$

$$= dx_k \wedge dy_k + (x_k \, dy_k + y_k \, dx_k) \wedge \frac{dr}{r},$$

and use the fact that $1 = \sum (x_k + iy_k)^2 = \sum(x_k^2 - y_k^2) + 2i \sum x_k y_k$, so $\sum d(x_k y_k) = \sum(x_k \, dy_k + y_k \, dx_k)$ vanishes on the quadric. Overall, we get

$$\sum dp_k \wedge dq_k = -\sum dx_k \wedge dy_k.$$

Note that this shows more generally that $T^* S^n$ is symplectomorphic to a smooth affine quadric in $n + 1$ complex variables. $\quad\square$

9

Surgery

In this chapter, we describe almost toric pictures of some of the most important surgery operations in four-dimensional topology. We first revisit blow-up but allow ourselves to blow up at an *edge point* on the toric boundary rather than a vertex. Then we discuss rational blow-up/blow-down. Finally, we use these ideas to explore the symplectic fillings of lens spaces.

9.1 Non-Toric Blow-Up

Let X be an almost toric manifold and let $x \in X$ be a toric fixed point, that is, a point lying over a Delzant vertex b in the almost toric base diagram. We have seen (Example 4.23) that performing the symplectic cut corresponding to truncating the moment polygon at b has the result of symplectically blowing up X at x. This is often called *toric blow-up* because it can be understood purely in terms of toric geometry. But what if we want to blow up a point $x' \in X$ which does not live over a Delzant vertex? In this section, we will explain how to blow up a point living in the toric boundary over an edge of the base diagram. We begin by describing the local picture; we use a strategy similar to what we used to analyse the Auroux system (Example 7.1).

Example 9.1 Consider the manifold $O(-1)$ from Example 3.18. Recall that this is the variety

$$\{(z_1, z_2, [z_3 : z_4]) \in \mathbb{C}^2 \times \mathbb{CP}^1 : z_1 z_4 = z_2 z_3\}$$

and that this is the blow-up of $\mathbb{C}^2$ at the origin. Pick a real number $c > 0$ and let $X = O(-1) \setminus \{z_1 = c\}$. We will write down an almost toric fibration on X whose toric boundary is the cylinder $\{(\xi, 0, [1 : 0]) : \xi \neq c\}$ and which has

122

one focus–focus fibre. Namely, we take

$$H(z_1, z_2, [z_3 : z_4]) = \left(|z_1 - c|^2, \frac{1}{2}|z_2|^2 - \frac{|z_3|^2}{|z_3|^2 + |z_4|^2} \right).$$

The function H_2 satisfies $H_2 \geq -1$ with equality if and only if $z_2 = z_4 = 0$. The image of H is $\{(b_1, b_2) \in \mathbb{R}^2 : b_1 > 0, b_2 \geq -1\}$. The toric boundary is the cylinder $\{(z_1, 0, [1 : 0]) \, z_1 \neq c\}$ living over the bottom edge $b_2 = -1$, and the focus–focus fibre is $H^{-1}(c^2, 0)$ (see Figure 9.1).

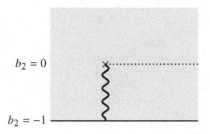

Figure 9.1 The local model for nontoric blow-up at a point over an edge in the toric boundary. The wiggly line indicates roughly where the exceptional sphere projects to. (Finding the precise image would require a nontrivial calculation of action coordinates.)

As with the Auroux system, the Hamiltonian H_2 generates a circle action, so the conversion to action coordinates has the form $(b_1, b_2) \mapsto (G_1(b_1, b_2), b_2)$. This means that (a) the toric boundary in action coordinates is still given by the horizontal line $b_2 = -1$, (b) the affine monodromy around the base-node is $\begin{pmatrix} 1 & 0 \\ 1 & 1 \end{pmatrix}$, so (c) both the toric boundary and the eigenline of the affine monodromy are horizontal, and choosing a horizontal branch cut, we get an almost toric base diagram of the form shown in Figure 9.1. The projection of the exceptional sphere $\{(0,0)\} \times \mathbb{CP}^1$ under H is a vertical line connecting the base-node to the toric boundary; I have not checked whether this remains vertical in action coordinates, so have drawn it as a wiggly line in Figure 9.1. Note that regardless of this projection, the symplectic area of the exceptional sphere is the affine length between the edge and the base-node.

Finally, we rotate the branch cut by 90° clockwise, to get the diagram shown in Figure 9.2. This picture can now be implanted near to an edge point in any almost toric base diagram; see Figure 9.3 for an example of a nine-point blow-up of $\mathbb{CP}^2$ (a *rational elliptic surface*).

Example 9.2 Consider the product symplectic manifold $X = T^2 \times \mathbb{C}$. We use 2π-periodic coordinates (θ_1, θ_2) on T^2 and Cartesian coordinates $x + iy$ on

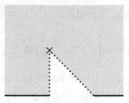

Figure 9.2 A different picture of the same local model as in Figure 9.1, obtained by rotating the branch cut 90° clockwise.

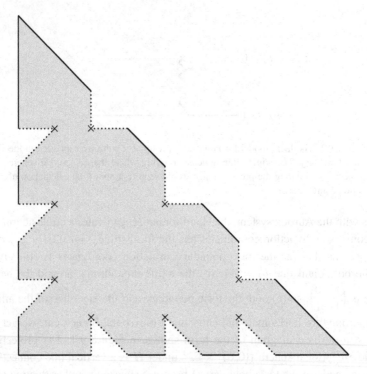

Figure 9.3 A non-toric blow-up of $\mathbb{CP}^2$ in nine balls.

$\mathbb{C}$, and equip X with the symplectic form $d\theta_1 \wedge d\theta_2 + dx \wedge dy$. The function $f: T^2 \times \mathbb{C} \to \mathbb{R}/2\pi\mathbb{Z} \times \mathbb{R}$ defined by $f(\theta_1, \theta_2, x + iy) = (\theta_1, (x^2 + y^2)/2)$ is a Lagrangian torus fibration, which induces the product integral affine structure on $(\mathbb{R}/2\pi\mathbb{Z}) \times \mathbb{R}$. The torus $T := T^2 \times \{0\}$ has self-intersection zero. We draw the integral affine base in Figure 9.4(a); the dotted lines with arrows indicate that the two sides of the picture should be identified.

Now perform non-toric blow-up at a point on T; we get the almost toric

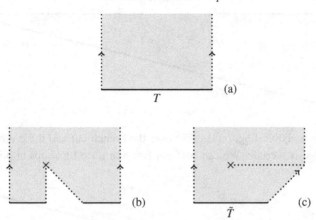

Figure 9.4 (a) The base of a Lagrangian torus fibration on $X = T^2 \times \mathbb{C}$.
(b) Non-toric blow-up of X. (c) Same diagram after a change of branch cut.

manifold whose base diagram is shown in Figure 9.4(b). Write E for the exceptional sphere and $\tilde{T}$ for the proper transform of T (which is the torus living over the toric boundary in Figure 9.4(b) or (c)). Since $\tilde{T} + E$ is homologous to a pushoff of T, we have $0 = (\tilde{T} + E)^2 = \tilde{T}^2 + 2\tilde{T} \cdot E + E^2$. Since $\tilde{T} \cdot E = 1$ and $E^2 = -1$, we have $\tilde{T}^2 = -1$. If we perform a change of branch cut on Figure 9.4(b), we get the diagram in Figure 9.4(c). By focusing on a neighbourhood of $\tilde{T}$, we get a local (almost toric) model for a symplectic manifold in a neighbourhood of a symplectic torus with self-intersection -1 (Figure 9.5(a)). Note that the matrix $\begin{pmatrix} 1 & 0 \\ 1 & 1 \end{pmatrix}$ is now used to identify the dotted edges with arrows. A similar argument gives a picture for tori with self-intersection $-n$ (Figure 9.5(b)).

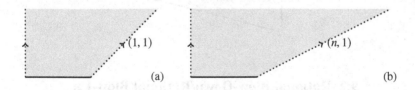

Figure 9.5 (a) Torus with self-intersection -1. (b) Torus with self-intersection $-n$.

Example 9.3 Consider the almost toric diagram of $\mathbb{CP}^2$ obtained from the standard diagram by a single nodal trade and mutation:

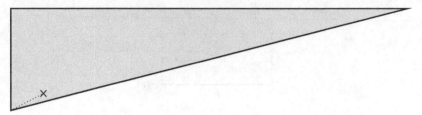

This has a visible Lagrangian $\mathbb{RP}^2$ over the branch cut and the preimage of the top edge is a conic. We can perform five non-toric blow-ups at points on the conic:

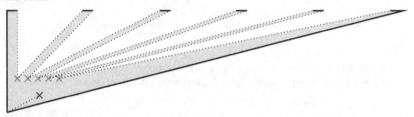

The 'bites' we have taken from the edge look a bit different to the usual picture in Figure 9.2, but are related to it by integral affine transformations. To be a little more precise, we take the affine length of the top edge to be 6 before the blow-up, and each bite takes out affine length 1 (so the corresponding exceptional sphere has area 2π). The result is a monotone[1] symplectic form on a Del Pezzo surface obtained by blowing up $\mathbb{CP}^2$ at five points. Changing branch cuts to make them parallel to their eigenlines, we obtain:

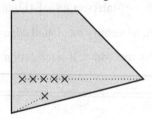

9.2 Rational Blow-Down/Rational Blow-Up

Recall that the lens space $L(n, a)$ is the quotient of the standard contact 3-sphere in $\mathbb{C}^2$ by the action of the cyclic group μ_n of nth roots of unity where μ acts by $(z_1, z_2) \mapsto (\mu z_1, \mu^a z_2)$. This lens space appears as the boundary of the symplectic orbifold $\mathbb{C}^2/\mu_n$, which is toric with moment polygon $\pi(n, a)$

---

[1] That is, the cohomology class of ω is a positive multiple of the first Chern class.

(Example 3.21). We can make symplectic cuts to get the *minimal resolution* of this orbifold singularity as in Example 4.30: we get a smooth symplectic 'filling' of $L(n, a)$ which retracts onto a chain of symplectic spheres whose self-intersection numbers $b_1, \ldots, b_k$ satisfy

$$\frac{n}{a} = b_1 - \cfrac{1}{b_2 - \cfrac{1}{\cdots - \frac{1}{b_k}}}.$$

Suppose that $n = p^2$ and $a = pq - 1$ for some coprime integers p, q. We have seen another symplectic filling of the lens space $L(p^2, pq - 1)$, namely $B_{1,p,q}$ from Example 7.8.

Definition 9.4   Suppose that $U \subseteq X$ is a symplectically embedded copy of (a neighbourhood of the chain of spheres in) the minimal resolution of $\frac{1}{p^2}(1, pq-1)$ inside a symplectic manifold X. The *rational blow-down*[2] of X along U is the symplectic manifold obtained by replacing U with (an open set in) $B_{1,p,q}$. Rational blow-up is the inverse operation.

Example 9.5   The toric diagram in Figure 9.6 shows a Hirzebruch surface containing a symplectic sphere (over the short edge) with self-intersection -4. We can rationally blow down along this sphere and we obtain an almost toric manifold containing a symplectically embedded $B_{1,2,1}$. In fact, this is symplectomorphic to $\mathbb{CP}^2$ (you can get back to the standard picture of $\mathbb{CP}^2$ by mutating).

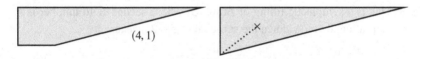

Figure 9.6 Rationally blow down the Hirzebruch surface $\mathbb{F}_4$ along a neighbourhood of the -4-sphere to get $\mathbb{CP}^2$ (the branch cut points in the $-(2, 1)$-direction).

Remark 9.6   Rational blow-up/blow-down was introduced by Fintushel and Stern [39], who showed that certain *elliptic surfaces* (minimal complex surfaces with Kodaira dimension 1) could be related by blow-ups and rational blow-downs, and used this to calculate their Donaldson invariants. Symington [105] used almost toric methods to show that these surgeries can be performed

---

[2]   Originally, rational blow-down was reserved for the case $q = 1$ and the more general procedure was called *generalised rational blow-down*. One could generalise still further using $B_{d,p,q}$ for $d \geq 2$, but the beauty of using $B_{1,p,q}$ is the drastic reduction in second Betti number that can be achieved.

symplectically. Rational blow-down has since been used extensively to construct exotic 4-manifolds with small second Betti number. This technique was pioneered by Park [84], who constructed a 4-manifold homeomorphic but not diffeomorphic to $\mathbb{CP}^2 \# 7\overline{\mathbb{CP}}^2$ by perfoming rational blow-down on a certain rational surface; the literature on 'small exotic 4-manifolds' has grown significantly since then.

Rather than focusing on exotica, we will content ourselves with constructing a symplectic filling of a lens space.

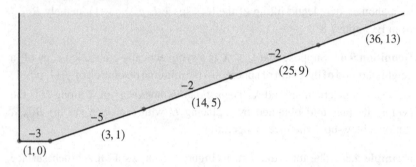

Figure 9.7 The minimal resolution of $\frac{1}{36}(1, 13)$. Delzant vertices are marked with dots and self-intersections of the spheres in the toric boundary are indicated.

Example 9.7 Consider the singularity $\frac{1}{36}(1, 13)$ and its minimal resolution, which contains a chain of spheres of self-intersections $-3, -5, -2, -2$; see Figure 9.8. This is a symplectic filling of $L(36, 13)$ whose second Betti number is 4. One can rationally blow down the sub-chain $-5, -2$, because

$$5 - \frac{1}{2} = \frac{9}{2},$$

which corresponds to $p^2/(pq - 1)$ for $p = 3$, $q = 1$. This gives a different symplectic filling of $L(36, 13)$ whose second Betti number is 1; see Figure 9.7.

Here is a different symplectic filling of $L(36, 13)$ which also has second Betti number equal to 1. First blow up the minimal resolution at the intersection point between the -3 and -5-spheres. This yields a chain of spheres with self-intersections $-4, -1, -6, -2, -2$. We can rationally blow down along both the -4-sphere and the $-6, -2, -2$-sub-chain, replacing them with $B_{1,2,1}$ and $B_{1,4,1}$ respectively (Figure 9.9).

Remark 9.8 (Exercise 9.19)   These two fillings are non-diffeomorphic: the first is simply-connected, while the second has fundamental group isomorphic to $\mathbb{Z}/2$.

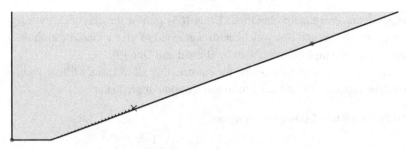

Figure 9.8 The result of rationally blowing down Figure 9.7 along the $-5, -2$ sub-chain of spheres; this contains a copy of $B_{1,3,1}$. The branch cut points in the $-(8, 3)$-direction (very close to the edge). The second Betti number of this filling is 1 (there is only one compact edge).

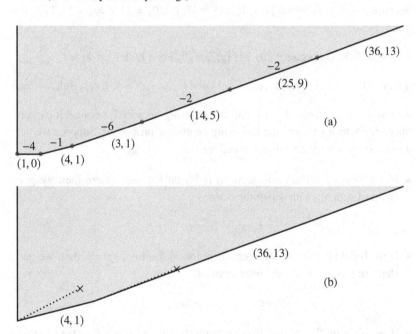

Figure 9.9 (a) Blow up the minimal resolution of $\frac{1}{36}(1, 13)$, (b) then rationally blow down along the -4 and $-6, -2, -2$ subchains. The branch cuts point in the $(2, 1)$ and $(8, 3)$-directions.

9.3 Symplectic Fillings of Lens Spaces

In Example 9.7, we constructed two non-diffeomorphic fillings of the lens space $L(36, 13)$ with second Betti number 1. In fact, symplectic fillings of lens spaces[3] are completely classified: Lisca [68] proved the classification up to diffeomorphism, and this was later strengthened to give a classification up to deformation/symplectomorphism by Bhupal and Ono [9].

There is an almost toric recipe for constructing all of Lisca's fillings, which we now explain. We first need to introduce some ingredients.

Definition 9.9 A continued fraction

$$[c_1, \ldots, c_m] = c_1 - \cfrac{1}{c_2 - \cfrac{1}{\cdots - \frac{1}{c_m}}}$$

is called a *zero continued fraction* (ZCF) if it evaluates to zero and its evaluation does not involve dividing by zero at any stage. For example, $[1, 1]$ is a ZCF because $1 - \frac{1}{1} = 0$, while $[1, 1, 1, 1, 1] = [1, 1, 1, 0] = [1, 1, \infty] = [1, 1] = 0$ is not.

Example 9.10 (Exercise 9.20) If $[c_1, \ldots, c_m]$ is a ZCF then so are

$$[1, c_1 + 1, \ldots, c_m], \quad [c_1, \ldots, c_m + 1, 1] \text{ and } [c_1, \ldots, c_i + 1, 1, c_{i+1} + 1, \ldots, c_m]$$

for any $i \in \{1, \ldots, m - 1\}$. We call this *blowing up* a ZCF because it captures the combinatorics behind the following geometric procedure. Suppose we have a chain of spheres with self-intersections $-c_1, \ldots, -c_m$:

- If we blow up a (non-intersection) point on the first sphere then we get a chain of spheres with self-intersections

$$-1, -c_1 - 1, -c_2, \ldots, -c_m.$$

- If we blow up a (non-intersection) point on the final sphere then we get a chain of spheres with self-intersections

$$-c_1, -c_2, \ldots, -c_m - 1, -1.$$

- If we blow up the intersection between the ith and $(i + 1)$st sphere then we get a chain of spheres with self-intersections

$$-c_1, \ldots, -c_i - 1, -1, -c_{i+1} - 1, \ldots, -c_m.$$

We call the obvious inverse procedure *blowing-down*.

Lemma 9.11 *Any ZCF is obtained from* $[1, 1]$ *by iterated blow-up.*

---

[3] There are many contact structures on lens spaces. We always equip $L(n, a)$ with its 'standard' contact structure descended from the tight contact structure on S^3.

Proof   Lemma 9.12 below implies that any ZCF of length at least 2 contains
an entry equal to 1, so can be blown down to get a shorter continued fraction.
Therefore, it is sufficient to prove that the only ZCF of length 2 is $[1, 1]$. If
$[c_1, c_2] = 0$ then $(c_1 c_2 - 1)/c_2 = 0$, so $c_1 c_2$ are positive integers whose product
is 1, hence $c_1 = c_2 = 1$.   □

Lemma 9.12 (Exercise 9.21)   *If $[c_1, \ldots, c_m]$ is a continued fraction with $c_i \geq 2$
for all i then it is not a ZCF. In fact, $[c_1, \ldots, c_m] > 1$.*

Corollary 9.13   *If $[c_1, \ldots, c_m]$ is a ZCF then there is a toric manifold containing a chain of spheres with self-intersections $-c_1, \ldots, -c_m$ such that the
moment polygon is an iterated truncation of Figure 9.10(a).*

Proof   Since $[c_1, \ldots, c_m]$ is obtained by blowing up $[1, 1]$, we simply follow
the geometric procedure outlined in Example 9.10 starting with Figure 9.10(a),
which contains a chain of spheres with self-intersections $-1, -1$. We can take
all the blow-ups to be toric (i.e. at vertices of the moment polygon).   □

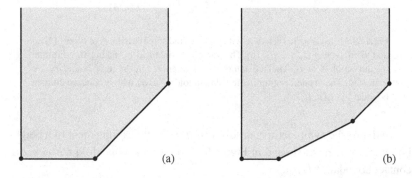

(a)   (b)

Figure 9.10 (a) Toric diagram corresponding to the ZCF $[1, 1]$. (b) Toric diagram
corresponding to a blow-up of $[1, 1]$ (in this case, $[2, 1, 2]$).

We need one final ingredient.

Example 9.14   The polygon in Figure 9.11(a) defines a toric variety X with
two cyclic quotient singularities A and B. The singularity A is $\frac{1}{n}(1, a)$; the
singularity B is isomorphic to $\frac{1}{n}(1, n - a)$: we can use the matrix[4] $\begin{pmatrix} -1 & -1 \\ 0 & 1 \end{pmatrix}$
to identify a neighbourhood of this vertex with a neighbourhood of the vertex
in $\pi(n, n - a)$. Let $\tilde{X}$ be the toric variety obtained by taking the minimal

---

[4] This matrix has determinant -1, which is responsible for the reversed ordering of the
exceptional spheres later.

resolution of X at B; this has the moment polygon shown in Figure 9.11(b). Note that if $\frac{n}{n-a} = [b_1, \ldots, b_m]$ then the self-intersections of the spheres in the minimal resolution appear as $-b_m, -b_{m-1}, \ldots, -b_1$ as we traverse the boundary anticlockwise.

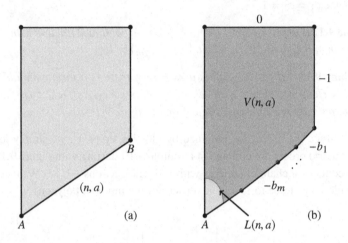

Figure 9.11 (a) A toric variety X with two orbifold singularities: A of type $\frac{1}{n}(1, a)$ and B of type $\frac{1}{n}(1, n - a)$. (b) The variety $\tilde{X}$ obtained by taking the minimal resolution of X at B. The self-intersections of curves in the toric boundary are indicated. The shaded region is the submanifold $V(n, a)$ whose concave contact boundary is $L(n, a)$.

We define the symplectic manifold $V(n, a)$ to be the complement of a neighbourhood of A in $\tilde{X}$, shaded in Figure 9.11(b). This submanifold has *concave* contact boundary[5] $L(n, a)$.

We will now give a recipe for constructing symplectic fillings of lens spaces.

Recipe 9.15   Let $[b_1, \ldots, b_m]$ be the continued fraction expansion of $n/(n-a)$. Suppose that[6] $[c_m, \ldots, c_1]$ is a ZCF with $c_i \leq b_i$ for $i = 1, \ldots, m$. Let Y be the corresponding toric variety from Corollary 9.13. If we perform $b_i - c_i$ non-toric blow-ups on the edge with self-intersection $-c_i$ then we obtain an almost toric manifold containing a chain of spheres with self-intersections $-b_m, \ldots, -b_1, -1, 0$. A neighbourhood of this chain is symplectomorphic to $V(n, a)$, so the complement of a neighbourhood of this chain gives a symplectic filling of $L(n, a)$.

---

[5] See Definition G.2.
[6] The reverse ordering is not a typo!

We illustrate this recipe with some simple examples.

Example 9.16 The fillings of $L(4, 1)$ were classified earlier by McDuff [74, Theorem 1.7]. Up to deformation, there are two: $B_{1,2,1}$ and $O(-4)$. We will construct these using Recipe 9.15. We have $\frac{4}{4-1} = [2, 2, 2]$. There are two possible ZCFs for use in Recipe 9.15: $[2, 1, 2]$ and $[1, 2, 1]$.

Case $[2, 1, 2]$: We start with the toric manifold shown in Figure 9.12(a) and perform a non-toric blow-up on the middle edge, yielding Figure 9.13(a). The red region is $V(4, 1)$ and its complement is a symplectic filling of $L(4, 1)$. If we perform a change of branch cut (Figure 9.14(a)), we see that this is precisely the almost toric diagram of $B_{1,2,1}$ from Example 7.8.

Case $[1, 2, 1]$: We start with the toric manifold shown in Figure 9.12(b) and perform a non-toric blow-up on the two outer edges, yielding Figure 9.13(b). The shaded region is $V(4, 1)$ and its complement is a symplectic filling of $L(4, 1)$. If we perform changes of branch cut (Figure 9.14(b)), we see that this is precisely the almost toric diagram of $O(-4)$ from Example 3.19.

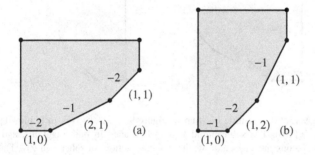

Figure 9.12 (a) The toric variety associated with the ZCF $[2, 1, 2]$. (b) The toric variety associated with the ZCF $[1, 2, 1]$.

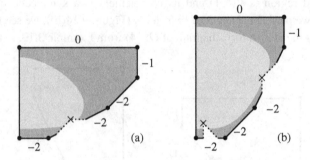

Figure 9.13 (a) Non-toric blow-up of Figure 9.12(a). (b) Non-toric blow-up of Figure 9.12(b). The submanifold $V(4, 1)$ is shaded in both diagrams, and the numbers along the edges indicate the self-intersections of spheres in $V(4, 1)$. The remaining grey regions are the fillings of $L(4, 1)$. In all cases, the eigenlines for the affine monodromy are parallel to the blown-up edges.

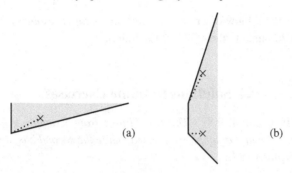

Figure 9.14 (a) The filling from Figure 9.13(a) after a change of branch cut; this
is $B_{1,2,1}$ from Example 7.8. (b) The filling from Figure 9.13(b) after a change of
branch cut; this is the standard toric picture of $O(-4)$ after two nodal trades and
a shear. In both cases we have also made deformations of the symplectic form to
shift the base-nodes around.

Theorem 9.17 (Lisca [68], Bhupal-Ono [9]) *Any symplectic filling of a lens
space is symplectomorphic to a deformation of a filling constructed using
Recipe 9.15.*

We will not prove this theorem, as it uses the theory of holomorphic curves
in a nontrivial way. Roughly, the idea is to cap off a symplectic filling U of
$L(n, a)$ using $V(n, a)$; the result contains an embedded symplectic sphere of
self-intersection 1 because $V(n, a)$ does (a smoothing of the 0 and -1-spheres
in the chain). By a result of McDuff [74, Theorem 1.1(i) + Theorem 1.4], this
implies that $U \cup V(n, a)$ is rational (an iterated symplectic blow-up of $\mathbb{CP}^2$),
so one reduces to studying different ways that $V(n, a)$ can embed in a rational
4-manifold.

Remark 9.18   If we remove all of the base-nodes from one of these almost toric
fillings by performing inverse generalised nodal trades, we obtain a (possibly)
singular toric variety with cyclic quotient T-singularities. This is a partial
resolution of $\frac{1}{n}(1, a)$. Note that Kollár and Shepherd-Barron used techniques
from Mori theory in [59] to show that any smoothing of $\frac{1}{n}(1, a)$ can be obtained
as a $\mathbb{Q}$-Gorenstein smoothing of a *P-resolution* of $\frac{1}{n}(1, a)$, that is, a partial
resolution with at worst T-singularities and such that all exceptional curves pair
nonnegatively with the canonical class. Stevens [104] and Christophersen [19]
showed that P-resolutions of $\frac{1}{n}(1, a)$ are in bijection with ZCFs $[c_1, \ldots, c_m]$
with $c_i \leq b_i$ where $n/(n - a) = [b_m, \ldots, b_1]$. The aforementioned prescription
gives us a way to go directly between the Lisca description of the filling and the
Kollár–Shepherd-Barron P-resolution. It also follows from [59, Lemma 3.14]
that all symplectic fillings of $L(n, a)$ are obtained by a combination of blow-

downs and rational blow-downs from a certain *maximal resolution* which is a P-resolution dominating the minimal resolution.

9.4 Solutions to Inline Exercises

Exercise 9.19 (Remark 9.8)   *The two fillings from Example 9.7 are non-diffeomorphic: the first is simply-connected, while the second has fundamental group isomorphic to $\mathbb{Z}/2$.*

Solution   We can deformation retract these almost toric manifolds onto the union of the almost toric boundary and the vanishing thimbles living over the branch cuts.

In the first case, the almost toric boundary is a sphere (over the compact edge) with two planes (over the noncompact edges) attached. We can ignore the planes as they are contractible. Thus the almost toric manifold deformation retracts onto a CW complex obtained by attaching a 2-cell to a sphere; the fundamental group is trivial by Van Kampen's theorem.

In the second case, the almost toric boundary is a cylinder C (with $\pi_1(C) = \mathbb{Z}$), and there are two vanishing thimbles. The boundaries of the vanishing thimbles attach to the cylinder in homotopy classes which correspond to the even numbers 2 and 4 in $\pi_1(C)$. You can see this because the vanishing thimbles are part of visible Lagrangian $(2, 1)$- and $(4, 1)$-pinwheels, so the thimble 'caps off' a loop which wraps twice (respectively four times) around the cylinder. By Van Kampen's theorem again, this means that the fundamental group of this CW complex is the quotient of $\mathbb{Z}$ by the subgroup generated by 2 and 4, that is $\mathbb{Z}/2$.   □

Exercise 9.20 (Example 9.10)   *If $[c_1, \ldots, c_m]$ is a zero continued fraction then so are $[1, c_1+1, \ldots, c_m]$, $[c_1, \ldots, c_m+1, 1]$ and $[c_1, \ldots, c_i+1, 1, c_{i+1}+1, \ldots, c_m]$ for any $i \in \{1, \ldots, m-1\}$.*

Solution   We deal with these three cases in order. Let $x = [c_2, \ldots, c_m]$. We have

$$[1, c_1 + 1, c_2, \ldots, c_m] = 1 - \frac{1}{c_1 + 1 - 1/x} = (c_1 - 1/x)/(c_1 + 1 - 1/x).$$

Since $c_1 - 1/x = 0$, this equals zero, proving the first case.

Blowing up at the end of a continued fraction does not change its value:

$$c_m = c_m + 1 - \frac{1}{1},$$

proving the second case.

Finally, we blow up in the middle of the chain. Let $x = [c_{i+2}, \ldots, c_m]$. We need to show that

$$c_i + 1 - \frac{1}{1 - \frac{1}{c_{i+1}+1-1/x}} = c_i - \frac{1}{c_{i+1} - 1/x}.$$

We have

$$c_i + 1 - \frac{1}{1 - \frac{1}{c_{i+1}+1-\frac{1}{x}}} = c_i + 1 - \frac{c_{i+1} + 1 - 1/x}{c_{i+1} - 1/x}$$

$$= c_i + \frac{c_{i+1} - 1/x - (c_{i+1} + 1 - 1/x)}{c_{i+1} - 1/x}$$

$$= c_i - \frac{1}{c_{i+1} - 1/x}.$$

This shows that blowing up the zero continued fraction does not change its value. $\square$

Exercise 9.21 (Lemma 9.12) *If $[c_1, \ldots, c_m]$ is a continued fraction with $c_i \geq 2$ for all i then it is not a ZCF. In fact, $[c_1, \ldots, c_m] > 1$.*

Proof We prove this by induction on the length of the continued fraction. It is clearly true if $m = 1$. Assume it is true for all continued fractions with length m and all entries ≥ 2; let $[c_1, \ldots, c_{m+1}]$ be a continued fraction of length $m + 1$. Then $[c_2, \ldots, c_m] > 1$ and $c_1 \geq 2$ by assumption, so

$$[c_1, \ldots, c_{m+1}] = c_1 - 1/[c_2, \ldots, c_{m+1}] > 2 - 1/1 = 1. \qquad \square$$

10

Elliptic and Cusp Singularities

I first learned about the following pictures in conversation with Paul Hacking, and then finally understood them by reading the paper [31] by Philip Engel.

10.1 Another Picture of $\mathbb{CP}^2$

Recall that to obtain an almost toric diagram for an almost toric fibration $f \colon X \to B$, we picked a simply-connected fundamental domain for the deck group action in the universal cover $\widetilde{B}^{reg}$. Since this fundamental domain is simply-connected, we can find single-valued action coordinates on the whole domain, and we took the almost toric diagram (fundamental action domain) to be the image of the fundamental domain under the action coordinates. In this section, we will allow ourselves something more exotic: we will take a branch cut in B whose complement is not simply-connected, but rather has fundamental group $\mathbb{Z}$. The action coordinates will be multi-valued, but related by a $\mathbb{Z}$-action which will be given by iterated application of an integral affine matrix. As a result, our pictures will have a high degree of redundancy (we are superimposing infinitely many fundamental action domains) but this can be fixed by quotienting them by the $\mathbb{Z}$-action. The result will be a 'conical' almost toric diagram rather than a planar diagram.

Example 10.1 Consider the standard almost toric picture of $\mathbb{CP}^2$ where we have made three nodal trades at the corners; there are three branch cuts extending from the nodes to the corners (Figure 10.1). The almost toric boundary is a cubic curve with self-intersection 9. We will redraw this picture as follows. Let A, B, C be the three nodes and let O be the barycentre of the triangle; let T be the tripod of lines OA, OB, OC (shown dashed in Figure 10.1). Let α, β, γ be the three triangular regions labelled in Figure 10.1.

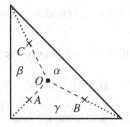

Figure 10.1 Changing branch cuts, reprise.

In Figure 10.2, we draw the image of the developing map on the complement of T in such a way that the almost toric boundary unwraps as a horizontal line. The image of the developing map is a strip (closed at the bottom, open at the top). We have shaded some fundamental action domains alternately light and dark. The fundamental group of the complement of T is $\mathbb{Z}$, which acts on the strip by powers of $\begin{pmatrix} 1 & 0 \\ 9 & 1 \end{pmatrix}$ (treating O as the origin). In each translate of the fundamental action domain, you can 'see' the hole left by excising T. For example, take the central light-coloured fundamental action domain. This is obtained by taking the region labelled γ in Figure 10.1 and appending the images of regions α and β under the monodromies around the nodes B and A respectively so that the almost toric boundary becomes straight.

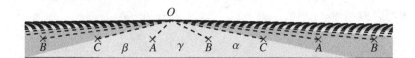

Figure 10.2 The image of the complement of T under the developing map for the integral affine structure; alternately shaded regions are fundamental action domains, tiling the strip. The dashed lines show where T has been excised.

Remark 10.2 (Exercise 10.21) The significance of the matrix $\begin{pmatrix} 1 & 0 \\ 9 & 1 \end{pmatrix}$ is that it is the total monodromy (anticlockwise) around the boundary loop in Figure 10.1, considered as starting in region γ. Note that 9 is also the self-intersection of the almost toric boundary curve. This is not a coincidence: compare with Example 9.2.

We will now be more explicit about monodromies. Let

$$M_A = \begin{pmatrix} 2 & 1 \\ -1 & 0 \end{pmatrix}, \quad M_B = \begin{pmatrix} -1 & 1 \\ -4 & 3 \end{pmatrix}, \quad M_C = \begin{pmatrix} -1 & 4 \\ -1 & 3 \end{pmatrix}$$

be the anticlockwise monodromies around the nodes A, B, C in Figure 10.1 (we can calculate these using[1] Lemma 6.13). Let $T^o := T \setminus \{0\}$ and call the three components 'branch cuts' (for now we ignore the interesting point 0). A path which crosses a branch cut lifts to a path in the strip but, as usual, the affine structure is twisted by a monodromy matrix when you cross the branch cut. In Figure 10.3, we label the branch cuts by their monodromies; these are obtained by conjugating the matrices M_A, M_B, M_C. For example, the nodes in the region γ are identical to A and B in Figure 10.1, so they have monodromies M_A and M_B. The node just to the right of B in Figure 10.3 is related to node C in Figure 10.1 by crossing the branch cut B anticlockwise, so its monodromy is $M_B M_C M_B^{-1}$. The next node to the right is related to node A by crossing the branch cuts B and then C anticlockwise, so its monodromy is $M_B M_C M_A M_C^{-1} M_B^{-1}$.

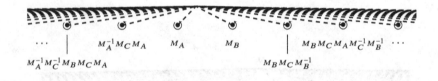

$$\cdots \quad | \quad M_A^{-1} M_C M_A \quad M_A \quad M_B \quad | \quad M_B M_C M_A M_C^{-1} M_B^{-1} \cdots$$

$$M_A^{-1} M_C^{-1} M_B M_C M_A \qquad\qquad M_B M_C M_B^{-1}$$

Figure 10.3 Monodromies for Figure 10.2 (all acting from the right).

We now try to understand what is happening near the point O. If we look at Figure 10.1, the point O looks less intimidating: it is just a point in B^{reg} away from the branch cuts. The integral affine structure around O is easy to understand: it is just an open ball in $\mathbb{R}^2$. The only reason it looks so interesting in Figure 10.2 is that the three legs of T pass through O. It becomes important to ask: if we have a diagram like Figure 10.2, how can we tell if the point O is a perfectly innocuous point in disguise? To answer this question, we first introduce the language of integral affine cones.

---

[1] Warning: the lemma tells us the *clockwise* monodromy.

10.2 Integral Affine Cones

Definition 10.3 Let $M \in SL(2, \mathbb{Z})$ and let $\langle M \rangle$ be the subgroup of $SL(2, \mathbb{Z})$ generated by M. Let $\ell \subseteq \mathbb{R}^2$ be a ray emanating from the origin. Consider the wedges $W_{M,\ell} \subseteq \mathbb{R}^2$ (respectively $W'_{M,\ell} \subseteq \mathbb{R}^2$) which are swept out[2] as $\ell M \neq \ell$ rotates anticlockwise (respectively clockwise) back to ℓ. We can equip the quotient space $B_{M,\ell} = W_{M,\ell} \setminus \{0\} / \langle M \rangle$ (respectively $B'_{M,\ell}$) with the structure of an integral affine manifold, by identifying $x \in \ell M$ with $xM^{-1} \in \ell$ at the level of points and $v \in T_x \mathbb{R}^2$ with $vM^{-1} \in T_{xM^{-1}} \mathbb{R}^2$ at the level of tangent vectors. We call such an integral affine manifold a *punctured cone*; we obtain a singular integral affine manifold called a *cone* by adding in the cone point 0. The matrix M is the *affine monodromy around the cone point*.

Remark 10.4 We can visualise $B_{M,\ell}$ as a cone obtained by wrapping $W_{M,\ell}$ up so that ℓ and ℓM are identified by the map $x \mapsto xM$; see Figure 10.4.

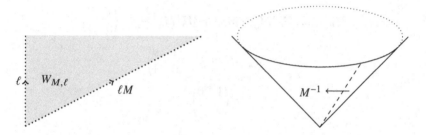

Figure 10.4 The cone $B_{M,\ell}$ obtained by 'wrapping up' $W_{M,\ell}$ using M to glue across the (dashed) branch cut.

Remark 10.5 Note that $W'_{M,\ell} = W_{M^{-1},\ell M}$ and $B'_{M,\ell} = B_{M^{-1},\ell M}$.

Exercise 10.6 *Make a paper model of the integral affine manifold $B_{M,\ell}$ for*

$$M = \begin{pmatrix} 0 & 1 \\ -1 & 0 \end{pmatrix}$$

and ℓ the positive x-axis. Is it possible to make paper models for $B_{M,\ell}$ when M is not conjugate to this matrix?

Lemma 10.7 *The integral affine structure on the punctured cone extends over the cone point when $M = I$.*

Proof In this case, $W_{M,\ell} = \mathbb{R}^2 \setminus \ell$ and we obtain the cone by identifying both sides of the cut using the identity, so the cone is just $\mathbb{R}^2$. □

---

[2] If $\ell M = \ell$ then both wedges are the whole of $\mathbb{R}^2$.

10.3 Back to the Example

Now take the horizontal line shown crossing the branch cuts in the top part of Figure 10.5. The portion of the almost toric base lying above this line is an integral affine cone: it is made up of infinitely many triangular segments which are glued together using the affine monodromies. To check that the integral affine structure extends over the cone point O, it suffices to check that the total monodromy as we traverse the boundary of this cone is the identity.

This horizontal line projects to a triangle in the lower part of Figure 10.5 (with corners where it hits the branch cuts). Since this triangle does not cross any of the dotted branch cuts, the monodromy around it is the identity, but we can perform the same calculation 'upstairs' by multiplying the clockwise monodromies of the three branch cuts crossed by the horizontal line, and then multiplying by $\begin{pmatrix} 1 & 0 \\ 9 & 1 \end{pmatrix}$ to send the lifted end-point back to the lifted start-point:

$$M_B^{-1} \cdot M_B M_C^{-1} M_B^{-1} \cdot M_B M_C M_A^{-1} M_C^{-1} M_B^{-1} \cdot \begin{pmatrix} 1 & 9 \\ 0 & 1 \end{pmatrix}$$

$$= \cdot M_A^{-1} M_C^{-1} M_B^{-1} \cdot \begin{pmatrix} 1 & 9 \\ 0 & 1 \end{pmatrix}$$

$$= \begin{pmatrix} 1 & 0 \\ 0 & 1 \end{pmatrix}.$$

10.4 Developing Map for Cones

We now discuss the developing map for the integral affine structure, and the dependence of $B_{M,\ell}$ on ℓ. Let $\tilde{B}_{M,\ell}$ be the universal cover of $B_{M,\ell}$. This can be constructed as follows. Take infinitely many copies W_i, $i \in \mathbb{Z}$, of $W_{M,\ell}$ and write $\partial_\ell W_i$ and $\partial_{\ell M} W_i$ for the two boundary rays. Let $\tilde{B}_{M,\ell}$ be the quotient of $\bigsqcup_{i \in \mathbb{Z}} W_i$ by identifying $x \in \partial_\ell W_i$ with $xM \in \partial_{\ell M} W_{i-1}$ for all $i \in \mathbb{Z}$. The image of W_i under the developing map for the integral affine structure on $\tilde{B}_{M,\ell}$ is then $M^i W_{M,\ell}$.

Lemma 10.8 *If $\ell' \subseteq M^i W_{M,\ell}$ for some i then $B_{M,\ell}$ and $B_{M,\ell'}$ are isomorphic as integral affine manifolds.*

Proof If $\ell' = \ell M^i$ then the map $M^i \colon W_{M,\ell} \to W_{M,\ell'}$ descends to give an isomorphism $B_{M,\ell} \to B_{M,\ell'}$. Using this, we may assume that $\ell' \subseteq W_{M,\ell}$.

Let S be the sector in $W_{M,\ell}$ swept out as ℓ' moves anticlockwise to ℓ and T

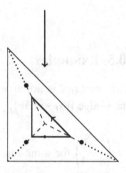

Figure 10.5 Bottom: A triangular path with trivial monodromy. Top: Its lift to the strip.

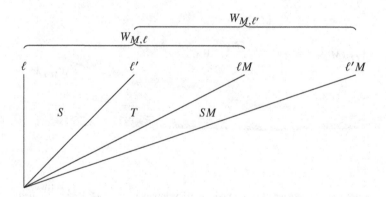

Figure 10.6 Constructing an isomorphism $B_{M,\ell} \to B_{M,\ell'}$. In this example, $M = \begin{pmatrix} 1 & 0 \\ 2 & 1 \end{pmatrix}$.

the other sector. We see that $W_{M,\ell'} = T \cup SM$. The piecewise linear map

$$W_{M,\ell} \to W_{M,\ell'}, \qquad x \mapsto \begin{cases} xM \text{ if } x \in S \\ x \text{ if } x \in T \end{cases}$$

descends to give the desired isomorphism $B_{M,\ell} \to B_{M,\ell'}$.   □

Lemma 10.9   *If $K \in GL(2, \mathbb{Z})$ then $B_{K^{-1}MK,\ell K}$ is isomorphic to $B_{M,\ell}$ via the map K. In particular, if $K = -I$ then we see that $B_{M,\ell} = B_{M,-\ell}$.*

Proof   If we consider K as a change of coordinates then, in the new coordinates, M is represented by $K^{-1}MK$ and ℓ is sent to ℓK. In particular, $W_{M,\ell}$ is sent to $W_{K^{-1}MK,\ell K}$, and the recipe for gluing together $\tilde{B}_{M,\ell}$ transforms into the recipe for gluing together $\tilde{B}_{K^{-1}MK,\ell K}$.   □

10.5 Examples

We will focus on examples where the ray ℓ is not an eigenray of M. We will also focus on examples where the wedge $W_{M,\ell}$ or $W'_{M,\ell}$ subtends an angle less then π radians.

Example 10.10   Consider $M = \begin{pmatrix} 1 & 0 \\ n & 1 \end{pmatrix}$ for some integer $n > 0$. If we take ℓ to be a ray pointing vertically up, then $W_{M,\ell}$ is shown in Figure 10.7 for $n = 3$. The image of $\tilde{B}_{M,\ell}$ is the strict upper half-plane, which is tiled by the domains $W_{M,\ell} M^i$.

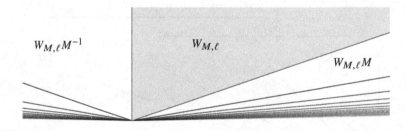

Figure 10.7 The image of $\tilde{B}_{M,\ell}$ under the developing map for $M = \begin{pmatrix} 1 & 0 \\ 2 & 1 \end{pmatrix}$, tiled by domains $W_{M,\ell} M^i$.

Any choice of ℓ in the strict upper half-plane will give the same $B_{M,\ell}$ by Lemma 10.8. If we take ℓ in the lower half-plane then we get the same integral affine manifold $B_{M,\ell}$ by Lemma 10.9.

Remark 10.11   The only other option is to take ℓ to be horizontal (an eigenray). If $n = 1$, the resulting $B_{M,\ell}$ is the integral affine base for a single focus–focus fibre.

Example 10.12 Consider[3] $M = \begin{pmatrix} 2 & 1 \\ 1 & 1 \end{pmatrix}$. This has eigenvalues $\frac{3\pm\sqrt{5}}{2}$ and eigenvectors $v_{\pm} = (x, y)$ with $y = -\frac{1}{2}(1 \mp \sqrt{5})x$. The eigendirections divide the plane into four open quadrants. Let us call the quadrants $Q, Q', -Q, -Q'$ where Q and Q' have the property that if $\ell \subseteq Q$ (respectively $\ell \subseteq Q'$) then $W_{M,\ell} \subseteq Q$ (respectively $W'_{M,\ell} \subseteq Q'$).

If we pick $\ell \subseteq Q$ or $\ell \subseteq -Q$ (respectively $\ell \subseteq Q'$ or $\ell \subseteq -Q'$) then the images $W_{M,\ell}M^i$ (respectively $W'_{M,\ell}M^i$), $i \in \mathbb{Z}$ tile the chosen quadrant. By Lemma 10.8, this means that, other than choosing an eigenray, the choice of ℓ amounts to a choice of quadrant. Moreover, rays from opposite quadrants yield the same integral affine manifold by Lemma 10.9. Thus, there are two *essentially different* choices of ray: $\ell_1 \subseteq Q$ or $\ell_2 \subseteq Q'$; see Figure 10.8. The same phenomenon occurs whenever M has two distinct positive real eigenvalues (which is equivalent to $\text{Tr}(M) \geq 3$).

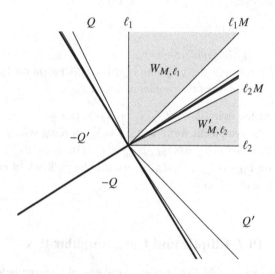

Figure 10.8 Some choices of rays ℓ_1 and ℓ_2 in Example 10.12 which give different manifolds B_{M,ℓ_1} and B'_{M,ℓ_2}. The thick lines are the eigenlines.

10.6 Symplectic Manifolds

Definition 10.13 We can associate to $B_{M,\ell}$ a symplectic manifold $X_{M,\ell}$ together with a Lagrangian torus fibration $X_{M,\ell} \to B_{M,\ell}$ compatible with the

---

[3] For 2-by-2 matrix connoisseurs, this is the Arnold cat map.

integral affine structure. To construct $X_{M,\ell}$, we take $W_{M,\ell} \times T^2$ with the symplectic form $\sum dp_i \wedge dq_i$ and identify $(\boldsymbol{p}, \boldsymbol{q})$ with[4] $(\boldsymbol{p}M^{-1}, M\boldsymbol{q})$ for $\boldsymbol{p} \in \ell M$.

Lemma 10.14 (Exercise 10.23)　*There exists a smooth path* $\gamma \colon [0,1] \to W_{M,\ell}$ *which is transverse to all rays emanating from the origin such that* $\gamma(0) \in \ell$, $\gamma(1) = \gamma(0)M$, *and* $\frac{d^n\gamma}{dt^n}(1) = \frac{d^n\gamma}{dt^n}(0)M$ *for all* $n \geq 1$.

Lemma 10.15　*Let* $Y_{M,\ell}$ *be the quotient of* $[0,1] \times T^2$ *which identifies* $(0, \boldsymbol{q})$ *with* $(1, M^T \boldsymbol{q})$. *This is a torus bundle over the circle. The symplectic manifold* $X_{M,\ell}$ *is diffeomorphic to* $(0, \infty) \times Y_{M,\ell}$.

Proof　Let $\gamma \colon [0,1] \to W_{M,\ell}$ be a path given by Lemma 10.14. This descends to a closed smooth loop in $B_{M,\ell}$. Moreover, because γ is transverse to all rays through the origin, we can foliate $B_{M,\ell}$ by loops $s\gamma$ for $s \in (0, \infty)$. Define a map $(0, \infty) \times [0,1] \times T^2 \to W_{M,\ell} \times T^2$ by $(s, t, \boldsymbol{q}) \mapsto (s\gamma(t), \boldsymbol{q})$. This descends to give a diffeomorphism $(0, \infty) \times Y_{M,\ell} \to X_{M,\ell}$. □

In fact, we can be more precise.

Lemma 10.16　$X_{M,\ell}$ *is symplectomorphic to* $(0, \infty) \times Y_{M,\ell}$ *with the symplectic form* $\omega = d(s\alpha)$, *where* $s \in (0, \infty)$ *and* α *is a contact form on* $Y_{M,\ell}$. *This is called the* symplectisation *of the contact manifold* $Y_{M,\ell}$.

Proof　Consider the map $(0, \infty) \times Y_{M,\ell} \to X_{M,\ell}$, $(s, t, \boldsymbol{q}) \mapsto (s\gamma(t), \boldsymbol{q})$ from Lemma 10.15. The symplectic form on $X_{M,\ell}$ is $\sum dp_i \wedge dq_i$ with $\boldsymbol{p} = s\gamma(t)$, so the pullback of this form to $(0, \infty) \times Y_{M,\ell}$ is $\sum_i d(s\gamma_i(t)) \wedge dq_i = d\left(\sum_i s\gamma_i(t)\, dq_i\right)$. The 1-form α on $Y_{M,\ell}$ is $\sum_i \gamma_i(t)\, dq_i$, so we see the pullback of ω along our diffeomorphism is $d(s\alpha)$ on $(0, \infty) \times Y_{M,\ell}$. □

10.7　Elliptic and Cusp Singularities

The integral affine manifold $B_{M,\ell}$ has a natural partial compactification,

$$\overline{B}_{M,\ell} := W_{M,\ell}/\langle M \rangle,$$

which adds in the cone point $\overline{b}$. If M is the identity matrix, we observed in Lemma 10.7 that the integral affine structure on $B_{M,\ell}$ extends over the cone point, and so $X_{M,\ell}$ naturally sits inside a larger symplectic manifold $\overline{X}_{M,\ell}$ which fibres over $\overline{B}_{M,\ell}$ with a regular torus fibre over $\overline{b}$.

If M is not the identity, we simply take $\overline{X}_{M,\ell}$ to be the partial compactification of $X_{M,\ell}$ which adds in a single point $\overline{x}$ over $\overline{b}$. We will think of this as a 'singular

---

[4] Compare with the map Φ in the proof of Lemma 2.25.

symplectic manifold', that is, a singular space with a symplectic form defined away from the singularity. In fact, this singular space can be equipped with the structure of a complex algebraic variety, with a singularity at $\bar{x}$ called an *elliptic* or *cusp singularity* depending on the matrix M; see Hirzebruch's seminal paper [54, Section 2.2–3]. The contact manifold $Y_{M,\ell}$ from Lemma 10.16 is known as the *link* of the singularity.

The minimal resolution of this algebraic variety replaces $\bar{x}$ with either a smooth elliptic curve (in the elliptic case), a nodal elliptic curve, or a cycle of rational curves. We will discuss the symplectic version of this, where we resolve the singularity by symplectic cuts.

Example 10.17 (Parabolic matrix)   Let $M = \begin{pmatrix} 1 & 0 \\ n & 1 \end{pmatrix}$ and take ℓ to point in the

$(0, 1)$-direction, as in Example 10.10. If we perform a symplectic cut on $\overline{X}_{M,\ell}$ at some positive height then we obtain the following integral affine base:

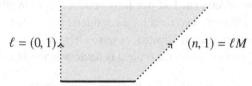

$$\ell = (0, 1) \qquad (n, 1) = \ell M$$

where we are identifying the edges labelled with arrows using the matrix M. The corresponding symplectic manifold is smooth, and lying over the compact edge we have a symplectic torus with self-intersection $-n$ (compare with Example 9.2). This corresponds to the case of an elliptic singularity, where the minimal resolution introduces a smooth elliptic curve.

Example 10.18 (Hyperbolic matrix)   Let $M = \begin{pmatrix} 2 & 1 \\ 1 & 1 \end{pmatrix}$ and take ℓ to point in the $(0, 1)$-direction as in Example 10.12. If we make a symplectic cut at some positive height then we obtain the following integral affine base:

$$\ell = (0, 1) \qquad (1, 1) = \ell M$$

where the edges with arrows are identified using M. The point[5] we have marked with a dot is actually a Delzant vertex. To see this, imagine how the left- and right-pointing edges look from the point of view of the leftmost dot

---

[5] Because of the identifications, there is only one point!

in the diagram. The right-pointing edge points in the $(1, 0)$-direction. The left-pointing edge points in the $(-1, 0)M^{-1} = (-1, 1)$-direction. We can make this clearer by shifting the branch cut ℓ to ℓ' parallel to $(-1, 3)$:

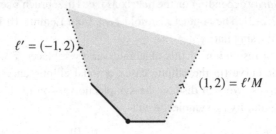

The toric boundary is therefore a nodal elliptic curve. This singularity is therefore a *cusp singularity*.

If this first symplectic cut had not fully resolved our singularity (had the corner not been Delzant), we could have continued in the manner of Example 4.30, making more cuts, until all vertices were Delzant. The result would be a cycle of rational curves; this is the typical behaviour when M is hyperbolic.[6]

Lemma 10.19 *If these rational curves have self-intersection* $s_1, s_2, \ldots, s_k$ *then the infinite periodic continued fraction*

$$s_1 - \cfrac{1}{s_2 - \cfrac{1}{\cdots - \cfrac{1}{s_k - \cfrac{1}{\cdots}}}}$$

converges to the slope of the dominant[7] *eigenline of* M.

Proof Suppose we make a sequence of cuts to the fundamental domain $W_{M,\ell}$ to get a new Delzant polygonal domain $\tilde{W}_{M,\ell}$. Using the action of $\langle M \rangle$, we get infinitely many translates of this domain.

If necessary, change the branch cut ℓ to ensure that none of the vertices of $\tilde{W}_{M,\ell}$ are on the branch cut. Let v_k be the primitive integer vectors pointing rightwards along the leftmost edge of $\tilde{W}_{M,\ell}M^k$. By construction, we have $v_k M = v_{k+1}$. The argument from Example 4.30 shows that $v_k S = v_{k+1}$ with $S = \begin{pmatrix} 0 & -1 \\ 1 & s_k \end{pmatrix} \cdots \begin{pmatrix} 0 & -1 \\ 1 & s_1 \end{pmatrix}$. Since $v_k = v_0 M^k$, we know that the slope of v_k converges to the slope of the dominant eigenline of M as $k \to \infty$. But as in Example 4.30, the recursion $v_k S = v_{k+1}$ tells us that if the slope of v_k is α_k

---

[6] That is, M has two distinct real eigenvalues.

[7] That is, the eigenline corresponding to the largest positive eigenvalue.

then the slope of v_{k+1} is

$$s_1 - \cfrac{1}{s_2 - \cfrac{1}{\cdots s_k - \frac{1}{\alpha_k}}}.$$

Therefore the infinite periodic continued fraction with coefficients

$$s_1, \ldots, s_k, s_1, \ldots, s_k, \ldots$$

converges to the slope of the dominant eigenline of M. □

Remark 10.20 Recall that for every hyperbolic matrix M there were two possible integral affine manifolds $B_{M,\ell}$ up to isomorphism, depending on the choice of ℓ. The corresponding cusp singularities are said to be *dual* to one another. Dual pairs of cusps were the subject of a long-standing conjecture of Looijenga, which was resolved by Gross, Hacking and Keel [48] using ideas from mirror symmetry. A different, more direct proof of this conjecture, which uses these almost toric pictures in an essential way was given by Engel [31].

10.8 K3 Surfaces from Fibre Sum

Take the almost toric picture of $\mathbb{CP}^2$ from Example 10.1; we redraw a single fundamental action domain in Figure 10.9(a), with boundary identifications indicated. The toric boundary is a symplectic torus with self-intersection 9, as we can see by comparing with Example 10.17. If we make nine non-toric blow-ups along the boundary as in Figure 9.3 then our picture changes: see Figure 10.9(b). We can make a change of branch cuts to make all of these branch cuts horizontal (Figure 10.9(c)); the opposite edges of the fundamental action domain are identified using the identity matrix or $\begin{pmatrix} 1 & 0 \\ 9 & 1 \end{pmatrix}$ depending on whether they are below (respectively above) the horizontal branch cut. The associated manifold $\mathbb{CP}^2 \sharp 9\overline{\mathbb{CP}}^2$ is called a *rational elliptic surface* and is often written $E(1)$ by low-dimensional topologists; its toric boundary is a torus with self-intersection 0.

The almost toric picture in Figure 10.10 is obtained by reflecting Figure 10.9(c) horizontally and ignoring the toric boundary. The associated almost toric manifold is obtained from a pair of rational elliptic surfaces by performing a *Gompf sum* [41] on the square-zero tori: in other words, we have excised a neighbourhood of the toric boundary in each copy of $E(1)$ and glued the results together along their common boundary T^3. This construction yields the elliptic surface $E(2)$, otherwise known as a K3 surface. The integral affine base is a sphere with 24 base-nodes: we have drawn a cylinder in Figure 10.10, and by Lemma 10.7, the integral affine structure extends over the sphere we get by adding in the points O and O' to the cylinder.

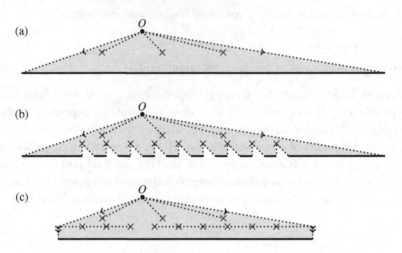

Figure 10.9 (a) The almost toric base diagram of $\mathbb{CP}^2$ from Example 10.1. The dotted edges are identified using the matrix $M = \begin{pmatrix} 1 & 0 \\ 9 & 1 \end{pmatrix}$. (b) Perform nine non-toric blow-ups along the toric boundary. (c) Make the branch cuts horizontal. The opposite edges are identified using M above the horizontal cut and using the identity below the cut.

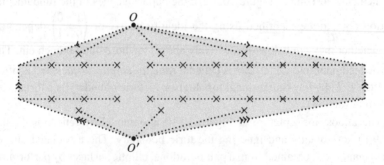

Figure 10.10 An almost toric fibration on a K3 surface. Edges are identified in pairs as indicated by the arrows.

10.9 Solutions to Inline Exercises

Exercise 10.21 (Remark 10.2) *Show that the matrix $\begin{pmatrix} 1 & 0 \\ 9 & 1 \end{pmatrix}$ is the total monodromy (anticlockwise) around the boundary loop in Figure 10.1 starting at a point in the region γ.*

Solution By Lemma 6.13, if the branch cut at a base-node points in the (p, q)-direction then the clockwise monodromy around that node is

$$\begin{pmatrix} 1 - pq & -q^2 \\ p^2 & 1 + pq \end{pmatrix}.$$

The branch cuts are in the directions $(-1, -1)$, $(2, -1)$, $(-1, 2)$ at A, B, and C respectively. This gives anticlockwise monodromy matrices

$$M_A = \begin{pmatrix} 2 & 1 \\ -1 & 0 \end{pmatrix}, \quad M_B = \begin{pmatrix} -1 & 1 \\ -4 & 3 \end{pmatrix}, \quad M_C = \begin{pmatrix} -1 & 4 \\ -1 & 3 \end{pmatrix}.$$

If we start with a vector v in the region γ then it crosses the branch cuts emanating from B, C, and A in that order, so the monodromy is

$$M_B M_C M_A = \begin{pmatrix} 1 & 0 \\ 9 & 1 \end{pmatrix}.$$

Note that if you start with a vector in the region β then you end up with the matrix $M_A M_B M_C$, which is different (though conjugate by M_A). □

Exercise 10.22 *Make a paper model of the integral affine manifold $B_{M,\ell}$ for*

$$M = \begin{pmatrix} 0 & 1 \\ -1 & 0 \end{pmatrix}$$

and ℓ the positive x-axis. Is it possible to make paper models for $B_{M,\ell}$ when M is not conjugate to this matrix?

Solution You can only make a paper model of the cone if M is an orthogonal (distance-preserving) map as well as being $\mathbb{Z}$-linear. The group $O(2) \cap SL(2, \mathbb{Z})$ consists of the four matrices $\begin{pmatrix} \pm 1 & 0 \\ 0 & \pm 1 \end{pmatrix}$, $\begin{pmatrix} 0 & \mp 1 \\ \pm 1 & 0 \end{pmatrix}$. These four are all possible to make. □

Exercise 10.23 (Lemma 10.14) *There exists a smooth path $\gamma \colon [0, 1] \to W_{M,\ell}$ which is transverse to all rays emanating from the origin such that $\gamma(0) \in \ell$, $\gamma(1) = \gamma(0)M$, and $\frac{d^n \gamma}{dt^n}(1) = \frac{d^n \gamma}{dt^n}(0)M$ for all $n \geq 1$.*

Solution   Let Θ be the angle between ℓ and ℓM subtended by $W_{M,\ell}$. Pick a segment of $\gamma\colon (-\epsilon, \epsilon) \to \mathbb{R}^2$ in a neighbourhood of ℓ so that $\gamma(0) \in \ell$ and so that γ is transverse to all rays emanating from the origin. In fact, we can assume by reparametrising γ that the ray through $\gamma(t)$ makes an angle Θt with ℓ for all $t \in (-\epsilon, \epsilon)$. Apply M to get a path-segment $\delta := \gamma M$ passing through ℓM. Again, by reparametrising, we can assume that $\delta(1) \in \ell M$ and $\delta(1+t)$ makes an angle Θt with ℓM for all $t \in (-\epsilon, \epsilon)$. We write these two segments in polar coordinates (r, θ) as $(r(t), \theta_0 + \Theta t)$ where θ_0 is the argument of the ray ℓ. We can now extend r in a completely arbitrary (smooth) way to the interval $[0, 1]$ and the graph will give a path with the desired properties. $\square$

Appendix A

Symplectic Linear Algebra

A.1 Symplectic Vector Spaces

Definition A.1   Let V be a finite-dimensional vector space over $\mathbb{R}$ and ω be a bilinear map $V \times V \to \mathbb{R}$. We say that ω is a *linear symplectic 2-form* if $\omega(v, v) = 0$ for all $v \in V$ and for all nonzero $v \in V$ there exists $w \in V$ such that $\omega(v, w) \neq 0$ (nondegeneracy). We say that the pair (V, ω) is a symplectic vector space.

Lemma A.2   *The map[1] $v \mapsto \iota_v \omega$ gives an isomorphism $V \to V^*$.*

Proof   If $\iota_v \omega = 0$ then $\omega(v, w) = 0$ for all w, so $v = 0$ by nondegeneracy. Therefore this map is an injective map between vector spaces of the same dimension, hence it is an isomorphism.   □

Definition A.3   If $W \subseteq V$ then we define the *symplectic orthogonal complement* $W^\omega \subseteq V$ to be the subspace

$$W^\omega := \{v \in V : \omega(v, w) = 0 \text{ for all } w \in W\}.$$

Lemma A.4 $\dim(W^\omega) = \dim(V) - \dim(W)$.

Proof   Under the isomorphism $v \mapsto \iota_v \omega$, the subspace W^ω is identified with the annihilator $W^\circ = \{f \in W^* : f(w) = 0 \text{ for all } w \in W\} \subseteq W^*$. The annihilator has dimension $\dim(V) - \dim(W)$ [53, §16, Theorem 1].   □

Lemma A.5   *Any nonzero finite-dimensional symplectic vector space V admits a basis $e_1, \ldots, e_n, f_1, \ldots, f_n$ such that $\omega(e_i, f_j) = \delta_{ij}$ and $\omega(e_i, e_j) = \omega(f_i, f_j) = 0$ for all i, j. Such a basis is called a* symplectic basis *for V. As a corollary, $\dim(V)$ is even.*

---

[1]   Recall that $\iota_v \omega$ denotes the 1-form $\omega(v, -)$.

Proof   Suppose that we have constructed a (possibly empty) partial symplectic basis $e_1, \ldots, e_k, f_1, \ldots, f_k$ of size $2k$, that is, a linearly independent set of $2k$ vectors satisfying the conditions in the statement of the lemma. Write $W \subseteq V$ for the span of this partial basis. Note that the restriction of ω to W is nondegenerate, so $W^\omega \cap W = 0$ and W^ω is a complement to W by Lemma A.4.

If $V \neq W$, pick $e_{k+1} \in W^\omega$. By construction, $\omega(e_{k+1}, e_i) = \omega(e_{k+1}, f_i) = 0$ for $i \leq k$. By nondegeneracy, we can find $f'_{k+1} \in W$ with $\omega(e_{k+1}, f'_{k+1}) = 1$. With respect to the splitting $V = W \oplus W^\omega$ we have $f'_{k+1} = g + f_{k+1}$ for uniquely determined vectors $g \in W$ and $f_{k+1} \in W^\omega$. Since $e_{k+1} \in W^\omega$, we have $\omega(e_{k+1}, g + f_{k+1}) = \omega(e_{k+1}, f_{k+1})$. Now $e_1, \ldots, e_{k+1}, f_1, \ldots, f_{k+1}$ is a partial symplectic basis of size $2(k + 1)$. At some point this construction terminates because V is finite-dimensional.   □

Definition A.6   Let (V, ω) be a symplectic vector space and $W \subseteq V$ a subspace. We say

- W is isotropic if $W \subseteq W^\omega$,
- W is coisotropic if $W^\omega \subseteq W$,
- W is Lagrangian if it is both isotropic and coisotropic,
- W is symplectic if $W \cap W^\omega = 0$.

Lemma A.7   *If W is isotropic then $2 \dim(W) \leq \dim(V)$. If W is coisotropic then $\dim(V) \leq 2 \dim(W)$. In particular, we see that Lagrangian subspaces satisfy $2 \dim(W) = \dim(V)$.*

Proof   If W is isotropic then we have

$$\dim(W) \leq \dim(W^\omega) = \dim(V) - \dim(W),$$

so $2 \dim(W) \leq \dim(V)$. If W is coisotropic then we have

$$\dim(V) - \dim(W) = \dim(W^\omega) \leq \dim(W),$$

so $\dim(V) \leq 2 \dim(W)$.   □

A.2   Complex Structures

Definition A.8   Let V be a vector space. A linear map $J : V \to V$ is called a *complex structure* if $J^2 = -I$.

Definition A.9   If (V, ω) is symplectic vector space then we say that:

- J tames ω if $\omega(v, Jv) > 0$ for any $v \neq 0$,
- J is ω-compatible if it tames ω and $\omega(Jv, Jw) = \omega(v, w)$ for all $v, w \in V$.

Lemma A.10 *Let (V, ω) be a symplectic vector space. If J is a complex structure on V taming ω and $W \subseteq V$ is a J-complex subspace (i.e. $JW = W$) then W is a symplectic subspace.*

Proof The subspace $W \cap W^\omega$ consists of vectors $v \in W$ such that $\omega(v, w) = 0$ for all $w \in W$. However, if $v \in W$ then $Jv \in W$, so $\omega(v, Jv) = 0$ and tameness implies $v = 0$. Thus $W \cap W^\omega = 0$ and W is symplectic. $\qquad\square$

Lemma A.11 *If (V, ω) is a symplectic vector space and J is an ω-compatible complex structure on V then $g_J(v, w) = \omega(v, Jw)$ defines a positive-definite symmetric bilinear form on V.*

Proof Bilinearity follows from bilinearity of ω. Symmetry follows from

$$g(w, v) = \omega(w, Jv) = \omega(Jw, J^2v) = \omega(Jw, -v) = \omega(v, Jw) = g(v, w).$$

Positive-definiteness follows from the fact that $g(v, v) = \omega(v, Jv) > 0$ if $v \neq 0$. $\qquad\square$

Lemma A.12 *Let J be an ω-compatible complex structure on a symplectic vector space (V, ω) and let W be a subspace. We have $W^\omega = (JW)^\perp$, where $\perp$ denotes the orthogonal complement with respect to g_J.*

Proof Since $g_J(v, Jw) = -\omega(v, w)$ we have

$$
\begin{aligned}
W^\omega &= \{v \in V : \omega(v, w) = 0 \text{ for all } w \in W\} \\
&= \{v \in V : g_J(v, Jw) = 0 \text{ for all } w \in W\} \\
&= (JW)^\perp.
\end{aligned}
$$
$\qquad\square$

Lemma A.13 *Let J be an ω-compatible complex structure on a symplectic vector space (V, ω) and let $L \subseteq V$ be a subspace. The following are equivalent.*

(a) L is Lagrangian;
(b) $L \perp JL$;
(c) JL is Lagrangian.

Proof We have $L^\omega = (JL)^\perp$, so $L = L^\omega$ if and only if $L \perp JL$. Thus (a) is equivalent to (b). Since $J^2 = -I$, $J^2L = L$, which means that (b) is symmetric in L and JL. Thus (b) is equivalent to (c). $\qquad\square$

Appendix B

Lie Derivatives

The background we assume on differential geometry can be found in many places, for example Lee's compendious book on smooth manifolds [63], Warner's terser book on manifolds and Lie theory [118], or Arnold's wonderful introduction to differential forms [3, Chapter 7]. However, the theory of Lie derivatives can be difficult to swallow on a first encounter. The philosophy behind this book has been to give a complementary perspective rather than rehashing what can be found written better elsewhere. In this appendix we give a quick and high-level conceptual overview of Lie derivatives from the point of view of Lie groups and Lie algebras, in the hopes that the reader will find this viewpoint helpful alongside a more traditional treatment. The final goal is to give a proof of the 'magic formulas' relating the Lie derivative, interior product, and exterior derivative.

B.1 Recap on Lie Groups

We start by giving a lightning review of Lie groups (see [118] for a more thorough introduction from the ground up). A *Lie group* G is a finite-dimensional manifold which is also a group in such a way that the multiplication and inversion maps are smooth. You will lose nothing by imagining that it is a group of matrices with real entries. The Lie algebra $\mathfrak{g}$ of G is the tangent space of G at the identity; in other words, if ϕ_t is a smooth path in G with $\phi_0 = \mathrm{id}$ then $\frac{d\phi_t}{dt}\big|_{t=0}$ is an element of $\mathfrak{g}$. Going back the other way, each $V \in \mathfrak{g}$ arises as the tangent vector $\frac{d\phi_t}{dt}\big|_{t=0}$ of a unique *1-parameter subgroup* ϕ_t of G, namely a path satisfying $\phi_0 = \mathrm{id}$ and $\phi_{s+t} = \phi_s \phi_t$ for all s, t. This 1-parameter subgroup is usually written as $\exp(tV)$.

The Lie algebra $\mathfrak{g}$ is a much simpler object than G: it is a vector space instead

156

of a manifold, so it has no interesting topology. It retains some knowledge of the group structure of G: it is equipped with an antisymmetric bilinear operation $[,]$ called *Lie bracket* which is defined as follows. If $V = \frac{d\psi_s}{ds}\big|_{s=0}$ and $W = \frac{d\phi_t}{dt}\big|_{t=0}$ then

$$[V, W] := \frac{d}{ds}\bigg|_{s=0} \frac{d}{dt}\bigg|_{t=0} \psi_s \phi_t \psi_s^{-1}.$$

If ψ_s and ϕ_t commute for all s, t then clearly $[V, W] = 0$. Somewhat miraculously, a kind of converse holds: if $[V, W] = 0$ then $\exp(sV)$ and $\exp(tW)$ commute. Indeed, the full multiplication structure of G in a neighbourhood of the identity can be determined from the Lie bracket.

The easiest example of a Lie group is the group $GL(n, \mathbb{R})$ of invertible n-by-n real matrices; indeed, its subgroups can be understood in the same way. Its Lie algebra is the space $\mathfrak{gl}(n, \mathbb{R})$ of all n-by-n real matrices, and the 1-parameter subgroup associated to a matrix V is

$$\exp(tV) = \sum_{n=0}^{\infty} \frac{1}{n!} V^n.$$

The Lie bracket is simply the commutator $[V, W] = VW - WV$.

One good way to understand a given Lie group G is to map it (smoothly and homomorphically) to a subgroup of $GL(n, \mathbb{R})$. Such a smooth homomorphism $R: G \to GL(n, \mathbb{R})$ is called a *representation* of G. Its differential at the identity matrix is a linear map

$$\rho := d_1 R \colon \mathfrak{g} \to \mathfrak{gl}(n, \mathbb{R})$$

which is a *representation* of the Lie algebra, in the sense that

$$[\rho(V), \rho(W)] = \rho([V, W]) \text{ for all } V, W \in \mathfrak{g}.$$

For example, $GL(n, \mathbb{R})$ acts on $\mathfrak{gl}(n, \mathbb{R})$ by conjugation, which gives a representation

$$\mathrm{Ad} \colon GL(n, \mathbb{R}) \to GL(\mathfrak{gl}(n, \mathbb{R})), \qquad \mathrm{Ad}(g)(V) = gVg^{-1}$$

and its differential is the representation

$$\mathrm{ad} \colon \mathfrak{gl}(n, \mathbb{R}) \to \mathfrak{gl}(\mathfrak{gl}(n, \mathbb{R})), \qquad \mathrm{ad}(V)(W) = [V, W].$$

These are both called the *adjoint* representation.

B.2 Diffeomorphism Groups

Let M be a smooth manifold and $G = \mathrm{Diff}(M)$ be the group of diffeomorphisms $M \to M$. This is not a Lie group, because it is not finite-dimensional, but many of the same ideas apply.

If we take a point $p \in M$ and a 1-parameter family of diffeomorphisms ϕ_t then we get a path $\phi_t(p)$. This gives a tangent vector $\left.\frac{d\phi_t(p)}{dt}\right|_{t=0}$ at p and hence a vector field on M whose value at p is

$$V(p) = \left.\frac{d\phi_t(p)}{dt}\right|_{t=0}.$$

Conversely, given a vector field V on M we get a 1-parameter subgroup of $\mathrm{Diff}(M)$ given by the flow[1] ϕ_t of V, which is the family of diffeomorphisms satisfying

$$\frac{d}{dt}\phi_t(p) = V(\phi_t(p)) \text{ for all } t \in \mathbb{R},\ p \in M.$$

For this reason, we will think of the space $\mathrm{vect}(M)$ of vector fields on M as the Lie algebra of $\mathrm{Diff}(M)$. We will figure out the Lie bracket in a moment.

The easiest way to study $\mathrm{Diff}(M)$ and its Lie algebra is via its representations. It comes with a natural and plentiful supply. We will write $\Omega^k(M)$ for the space of smooth differential k-forms (so $\Omega^0(M)$ means smooth functions).

Example B.1 $\mathrm{Diff}(M)$ acts on functions by pullback:

$$f \mapsto \phi^* f := f \circ \phi.$$

This is a *right action* in the sense that $(\phi_1 \circ \phi_2)^* = \phi_2^* \circ \phi_1^*$ so it gives an *antirepresentation*[2]

$$R\colon \mathrm{Diff}(M) \to GL(\Omega^0(M)), \qquad \phi \mapsto \phi^*.$$

This therefore gives an antirepresentation $\rho\colon \mathrm{vect}(M) \to \mathfrak{gl}(\Omega^0(M))$ of the Lie algebra $\mathrm{vect}(M)$ by defining

$$\rho(V)(f) = \left.\frac{d}{dt}\right|_{t=0} \phi_t^* f$$

where ϕ_t is the flow of V. This is better known as the *directional derivative* of

[1] It is, of course, possible that the flow is only locally defined, or defined for small t. For example, if $M = \mathbb{R}$ and $V(x) = -x^2$ then the flow is $\phi_t(x) = x/(xt+1)$, and we see that $\lim_{t \to -1/x} \phi_t(x) = \infty$. We avoid this kind of behaviour if M is compact or if ϕ_t preserves the (compact) level sets of some proper function, for example, if ϕ_t is the Hamiltonian flow of a proper Hamiltonian.

[2] If the notation $GL(\Omega^0(M))$ is giving you a headache, it just means 'invertible linear maps $\Omega^0(M) \to \Omega^0(M)$'. Similarly, $\mathfrak{gl}(\Omega^0(M))$ will mean 'linear maps $\Omega^0(M) \to \Omega^0(M)$'.

f in the V-direction, and written $V(f)$. In fact, since pullback of functions is multiplicative

$$\phi^*(fg) = (\phi^* f)(\phi^* g),$$

the representation R lands in the subgroup $\mathrm{Aut}(\Omega^0(M)) \subseteq GL(\Omega^0(M))$ of automorphisms of the ring of functions, and so ρ lands in the subalgebra $\mathfrak{der}(\Omega^0(M)) \subseteq \mathfrak{gl}(\Omega^0(M))$ of derivations of the ring of functions. In fact, this representation is injective: we can really think of vector fields as derivations on functions without losing information, and many expositions *define* vectors as derivations.

We can now identify the Lie bracket on Vect(M): because ρ is an *anti*representation, it should be *minus* the commutator bracket on derivations. To avoid sign-clashes with the rest of the literature, we actually write $[,]$ for the commutator bracket

$$[V, W](f) = V(W(f)) - W(V(f))$$

and call it 'Lie bracket of vector fields', even though it is out by a minus sign.

Example B.2 Pullback of differential forms gives another natural antirepresentation

$$\mathrm{Diff}(M) \to GL(\Omega^*(M)), \qquad \phi \mapsto \phi^*.$$

This preserves wedge product of differential forms:

$$\phi^*(\eta_1 \wedge \eta_2) = \phi^* \eta_1 \wedge \phi^* \eta_2.$$

We write $V \mapsto \mathcal{L}_V$ for the corresponding Lie algebra antirepresentation $\mathrm{vect}(M) \to \mathfrak{gl}(\Omega^*(M))$:

$$\mathcal{L}_V \eta := \frac{d}{dt}\bigg|_{t=0} \phi_t^* \eta.$$

By the Leibniz rule, $\mathcal{L}_V$ acts as a *derivation* of the algebra $\Omega^*(M)$:

$$\mathcal{L}_V(\eta_1 \wedge \eta_2) = (\mathcal{L}_V \eta_1) \wedge \eta_2 + \eta_1 \wedge \mathcal{L}_V \eta_2, \qquad \mathcal{L}_V d\eta = d\mathcal{L}_V \eta.$$

Moreover, since this is an *anti*representation, and since we have grudgingly accepted a historical minus sign in our bracket, we get

$$\mathcal{L}_{[V,W]} \eta = \mathcal{L}_V \mathcal{L}_W \eta - \mathcal{L}_W \mathcal{L}_V \eta. \tag{B.1}$$

B.3 Cartan's Magic Formulas

The algebra of differential forms admits further natural operations which have played a key role in this book. Our goal here is to prove the 'magic formulas' which govern the interplay between these operations and $\mathcal{L}_V$.

The operations in question are the exterior derivative,[3]

$$d: \Omega^*(M) \to \Omega^{*+1}(M),$$

and the interior product

$$\iota_V: \Omega^*(M) \to \Omega^{*-1}(M), \qquad (\iota_V \eta)(V_1, \ldots, V_{k-1}) = \eta(V, V_1, \ldots, V_{k-1}),$$

where V is a choice of vector field. Both of these operations are *antiderivations*, that is,

$$d(\eta_1 \wedge \eta_2) = (d\eta_1) \wedge \eta_2 + (-1)^{|\eta_1|} d\eta_2.$$

First, we prove some lemmas.

Lemma B.3

$$d\mathcal{L}_V \eta = \mathcal{L}_V d\eta.$$

Proof Let ϕ_t be the flow along V. We have $\phi_t^* d\eta = d\phi_t^* \eta$. Differentiating with respect to t at $t = 0$ gives the required identity. □

Lemma B.4 *If f is a function (0-form) then*

$$\mathcal{L}_V f = \iota_V df.$$

Proof Let ϕ_t be the flow of V. In local coordinates, for small t, we have[4] $\phi_t(p) = p + tV(p) + o(t)$ and therefore

$$(\phi_t^* f)(p) = f(\phi_t(p)) = f(p) + t\, df(V(p)) + o(t).$$

This means that

$$\left.\frac{d}{dt}\right|_{t=0} (\phi_t^* f)(p) = df(V(p)),$$

or

$$\mathcal{L}_V f = \iota_V df. \qquad □$$

Theorem B.5 (Cartan's magic formulas) *We have*

$$\iota_V d + d\iota_V = \mathcal{L}_V, \tag{B.2}$$

$$\mathcal{L}_V \iota_W - \iota_W \mathcal{L}_V = \iota_{[V,W]}. \tag{B.3}$$

---

[3] The most conceptually well-motivated exposition of d that I know is due to Arnold: [3, Chapter 7].

[4] As usual, $o(t)$ denotes a quantity such that $\lim_{t \to 0} o(t)/t = 0$.

Proof The operators d and ι_V are antiderivations and $\mathcal{L}_V$ is a derivation. This implies that the operator $\iota_V d + d\iota_V$ is a derivation and that $\mathcal{L}_V \iota_W - \iota_W \mathcal{L}_V$ is an antiderivation. Derivations and antiderivations on $\Omega^k(M)$ are determined by their effect on functions and on exact 1-forms. This is easy to see in local coordinates: if

$$\eta = \sum \eta_{i_1 \cdots i_k} dx_{i_1} \wedge \cdots \wedge dx_{i_k}$$

and D is a derivation, for example, then

$$D\eta = \sum (D\eta_{i_1 \cdots i_k}) dx_{i_1} \wedge \cdots \wedge dx_{i_k} + \sum \eta_{i_1 \cdots i_k} (D(dx_{i_1})) \wedge \cdots \wedge dx_{i_k} +$$
$$+ \cdots + \sum \eta_{i_1 \cdots i_k} dx_{i_1} \wedge \cdots \wedge (D(dx_{i_k})),$$

so D is determined completely by its action on functions (like the coefficients $\eta_{i_1 \cdots i_k}$) and exact 1-forms (like the local coordinate 1-forms dx_i).

Therefore it suffices to check Equation (B.2) and (B.3) for functions and for exact 1-forms.

Equation (B.2) For functions f, Equation (B.2) is simply the identity $\mathcal{L}_V f = df(V)$ which we proved in Lemma B.4. Now suppose we have an exact 1-form df. We have $(d\iota_V + \iota_V d)df = d\iota_V df$ because $d^2 = 0$ and

$$\mathcal{L}_V df = d\mathcal{L}_V f = d\iota_V df,$$

using Lemmas B.3 and B.4, so both sides of Equation (B.2) agree when applied to df. This proves Equation (B.2).

Equation (B.3) For functions f, Equation (B.3) reduces to $0 = 0$. For exact 1-forms df, using $\iota_W df = \mathcal{L}_W f$, the left-hand side of the Equation (B.3) becomes

$$\mathcal{L}_V \mathcal{L}_W f - \iota_W \mathcal{L}_V df.$$

Since $\iota_W \mathcal{L}_V df = \iota_W d\mathcal{L}_V = \mathcal{L}_W \mathcal{L}_V f$, this becomes $(\mathcal{L}_V \iota_W - \iota_W \mathcal{L}_V) df = (\mathcal{L}_V \mathcal{L}_W - \mathcal{L}_W \mathcal{L}_V) f$, which becomes $\mathcal{L}_{[V,W]} f$ using Equation (B.1). Therefore

$$(\mathcal{L}_V \iota_W - \iota_W \mathcal{L}_V) df = \mathcal{L}_{[V,W]} f = \iota_{[V,W]} df.$$

This proves Equation (B.3) for exact 1-forms and hence in general. $\square$

Appendix C

Complex Projective Spaces

C.1 CP^n

Definition C.1 The complex projective space CP^n is the space of complex lines in $\mathbb{C}^{n+1}$ passing through the origin.

Recall that a complex line passing through the origin is a subspace of the form $\mathbb{C} \cdot z := \{(\lambda z_1, \ldots, \lambda z_{n+1}) : \lambda \in \mathbb{C}\} \subseteq \mathbb{C}^{n+1}$ for some complex vector $z = (z_1, \ldots, z_{n+1}) \neq 0$.

Lemma C.2 (Exercise C.16) *If $z, z' \neq 0$ are complex vectors then $\mathbb{C} \cdot z = \mathbb{C} \cdot z'$ if and only if $z = \mu z'$ for some complex number $\mu \neq 0$.*

Lemma C.3 *The complex projective space CP^n is the quotient of $\mathbb{C}^{n+1} \setminus \{0\}$ by the equivalence relation $z \sim z'$ if and only if $z' = \mu z$ for some complex number $\mu \neq 0$.*

Proof Each $z \in \mathbb{C}^{n+1} \setminus \{0\}$ gives us a line $\mathbb{C} \cdot z$ and every complex line has this form, so the map

$$Q: \mathbb{C}^{n+1} \setminus \{0\} \to CP^n, \qquad Q(z) = \mathbb{C} \cdot z$$

is a surjection. By Lemma C.2, the fibres of $\mathbb{C} \cdot$ are the stated equivalence classes. $\qquad \square$

We equip CP^n with the quotient topology induced by Q. We often write $[z]$ or $[z_1 : \cdots : z_{n+1}]$ for $\mathbb{C} \cdot z$, and call the z_i *homogeneous coordinates* on CP^n. Homogeneous coordinates are not like Cartesian coordinates: Cartesian coordinates have the property that if p and q have different coordinates then $p \neq q$, but the homogeneous coordinates $[1 : 1]$ and $[2 : 2]$ specify the same point in CP^1. We remedy this redundancy, at the cost of missing some points, by passing to *affine charts*.

162

Definition C.4 Let $A_k = \{(z_1, \ldots, z_{n+1}) \in \mathbb{C}^{n+1} : z_k = 1\} \subseteq \mathbb{C}^{n+1}$.

Lemma C.5 (Exercise C.17) *The restriction $Q|_{A_k} : A_k \to \mathbb{CP}^n$ is an embedding. We call its image an* affine chart *in $\mathbb{CP}^n$.*

Lemma C.6 *The topological space $\mathbb{CP}^n$ is a complex manifold (in fact, an algebraic variety).*

Proof If $z \in \mathbb{C}^{n+1}$ has $z_k \neq 0$ then $z/z_k \in A_k$ and $Q(z) = Q(z/z_k)$, so $Q(z) \in Q(A_k)$. The space $\mathbb{CP}^n$ is therefore covered by the $n+1$ affine charts $Q(A_1), \ldots, Q(A_{n+1})$. The transition map

$$\varphi_{k\ell} := Q|_{A_k}^{-1} \circ Q|_{A_\ell} : \{z \in A_\ell : z_k \neq 0\} \to \{z \in A_k : z_\ell \neq 0\}$$

is given by

$$\varphi_{k\ell}(z_1, \ldots, z_k, \ldots, z_\ell = 1, \ldots, z_{n+1}) = \left(\frac{z_1}{z_k}, \ldots, 1, \ldots, \frac{1}{z_k}, \ldots, \frac{z_{n+1}}{z_k}\right),$$

which is an algebraic isomorphism, so $\mathbb{CP}^n$ is an algebraic variety[1] locally modelled on $\mathbb{C}^n$ and hence a complex manifold. □

Example C.7 The complex projective 1-space $\mathbb{CP}^1$ is diffeomorphic to the 2-sphere. To see this, let (r, θ, h) be cylindrical polar coordinates on S^2 and define σ_1 and σ_2 by

$$\sigma_1(r, \theta, h) = \frac{r}{1-h} e^{i\theta}, \qquad \sigma_2(r, \theta, h) = \frac{r}{1+h} e^{i\theta}$$

for $h < 1$ and $h > -1$ respectively. These are the stereographic projections from the North and South poles respectively. Since $h^2 + r^2 = 1$, we have $\sigma_1 \overline{\sigma}_2 = 1$, so if we use σ_1 and $\overline{\sigma}_2$ as coordinate charts on S^2 then the transition map is $\sigma_1 \mapsto \overline{\sigma}_2 = 1/\sigma_1$. This is the same as the transition map $\varphi_{1,2}$ for $\mathbb{CP}^1$ from the proof of Lemma C.6, so these manifolds are diffeomorphic.

Lemma C.8 *The complex projective space $\mathbb{CP}^n$ is the quotient of the unit sphere $S^{2n+1} \subseteq \mathbb{C}^{n+1}$ by the $U(1)$-action where $u \in U(1)$ acts on $\mathbb{C}^{n+1}$ by $z \mapsto uz$. Here, $U(1)$ is the multiplicative group of unit complex numbers.*

Proof The restriction of Q to $S^{2n+1} \subseteq \mathbb{C}^{n+1} \setminus \{0\}$ is still surjective because every complex line contains a circle of vectors of unit length. The fibres of $Q|_{S^{2n+1}}$ are precisely these unit circles in each complex line, which are also the orbits of the $U(1)$-action in the statement of the lemma. □

---

[1] Just as one can define a manifold as a collection of charts glued together by transition maps, an algebraic variety can be defined as a collection of affine varieties glued together by algebraic transition maps. Affine varieties are just subsets of affine space cut out by polynomial equations. In this case, the affine charts are copies of $\mathbb{C}^n$ rather than something more exotic.

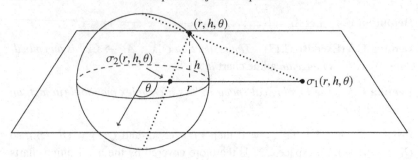

Figure C.1 Stereographic projections.

C.2 Projective Varieties

A polynomial $F(z_1, \ldots, z_{n+1})$ is called *homogeneous of degree d* if

$$F(\lambda z_1, \ldots, \lambda z_{n+1}) = \lambda^d F(z_1, \ldots, z_{n+1})$$

for all $\lambda \in \mathbb{C}$. For example, $z_1 z_2 + z_3 z_4$ is homogeneous of degree 2, whereas $z_1 + z_2^2$ is not homogeneous of any degree. It is not hard to show that a homogeneous polynomial of degree d is precisely a linear combination of monomials, each of which has degree precisely d.

The advantage of working with homogeneous polynomials is that if $F(z) = 0$ and $[z] = [z']$ then $F(z') = 0$. In other words, it makes sense to write $F([z]) = 0$: this condition doesn't depend on the choice of the homogeneous coordinate z.

Definition C.9 We define the *projective subvariety cut out by F*, to be the subset $\mathbb{V}(F) := \{[z] \in \mathbb{CP}^n : F([z]) = 0\}$. Similarly, we can define subvarieties cut out by a (possibly empty) set of homogeneous polynomials $\{F_1, \ldots, F_s\}$. Note that $\mathbb{V}(\emptyset) = \mathbb{CP}^n$.

Example C.10 Let $F(z_1, z_2) = z_1 z_2$. Then $\mathbb{V}(F) \subseteq \mathbb{CP}^1$ consists of the points $[1 : 0]$ and $[0 : 1]$.

Remark C.11 In affine algebraic geometry, an affine variety is the subset cut out of an affine space by a collection of (not necessarily homogeneous) polynomials. The intersection of the projective variety $\mathbb{V}(F)$ with the affine chart $Q(A_k)$ is defined by $F(z_1, \ldots, z_k = 1, \ldots, z_{n+1}) = 0$, which is a (not necessarily homogeneous) polynomial, so $\mathbb{V}(F) \cap Q(A_k)$ is an affine variety in the chart $Q(A_k)$.

Remark C.12 Unless F is constant, the expression $F([z])$ does not make sense as a complex-valued function on $\mathbb{CP}^n$. For example, if $F(z) \neq 0$ and $d \geq 1$ then $F(2z) = 2^d F(z) \neq F(z)$, but $[2z] = [z]$.

C.3 Zariski-Closure

We will use the notion of Zariski-closure in Chapter F.

Definition C.13 The *Zariski-closure* of a subset $S \subseteq \mathbb{CP}^n$ is the smallest subvariety containing S. If $V \subseteq \mathbb{CP}^n$ is a subvariety then a subset $S \subseteq V$ is called *Zariski-dense* in V if its Zariski-closure is V.

Example C.14 Consider the set of points $S = \{[n : 1] : n \in \mathbb{Z}\} \subseteq \mathbb{CP}^1$. The Zariski-closure of this set is $\mathbb{CP}^1 = \mathbb{V}(0)$. This is because if there is a homogeneous polynomial $F = \sum_{m=0}^{d} a_m z_1^m z_2^{d-m}$ with $F([n : 1]) = 0$ for all $n \in \mathbb{Z}$ then $\sum_{m=0}^{d} a_m z^m$ has infinitely many zeros (every integer) and hence vanishes identically.

Example C.15 Consider the set of points $S = \{[z : 1 : 0] \in \mathbb{CP}^2 : z \in \mathbb{C}\}$. This is contained in the subvariety $\mathbb{V}(z_3)$ defined by the vanishing of the z_3-coordinate. Moreover, it is Zariski-dense inside that subvariety. To see this, suppose that $F = \sum_{m_1+m_2+m_3=d} a_{m_1,m_2,m_3} z_1^{m_1} z_2^{m_2} z_3^{m_3}$ is a homogeneous polynomial with $F(z, 1, 0) = 0$ for all $z \in \mathbb{C}$. Then $\sum_{m_1,m_2} a_{m_1,m_2,0} z^{m_1} = 0$ has infinitely many solutions (any $z \in \mathbb{C}$), so $a_{m_1,m_2,0} = 0$ for all m_1, m_2. Thus, F is divisible by z_3. Therefore $\mathbb{V}(F)$ contains $\mathbb{V}(z_3)$. Thus, $\mathbb{V}(z_3)$ is the smallest subvariety containing S.

C.4 Solutions to Inline Exercises

Exercise C.16 (Lemma C.2) *If $z, z' \neq 0$ are complex vectors then $\mathbb{C} \cdot z = \mathbb{C} \cdot z'$ if and only if $z = \mu z'$ for some complex number $\mu \neq 0$.*

Solution If $z = \mu z'$ then $\mathbb{C} \cdot z = \{\lambda z \ \lambda \in \mathbb{C}\} = \{\lambda \mu z' : \lambda \in \mathbb{C}\}$ and as λ varies over $\mathbb{C}$, so $\lambda \mu$ varies over $\mathbb{C}$, so $\mathbb{C} \cdot z = \mathbb{C} \cdot z'$. Conversely, if $\mathbb{C} \cdot z = \mathbb{C} \cdot z'$ then $z' \in \mathbb{C} \cdot z$ so $z' = \mu z$ for some $\mu \in \mathbb{C}$. Since $z' \neq 0$, $\mu \neq 0$. □

Exercise C.17 (Lemma C.5) *The restriction $Q|_{A_k} : A_k \to \mathbb{CP}^n$ is an embedding.*

Proof If $z, z' \in A_k$ then $z_k = z'_k = 1$. If $Q(z) = Q(z')$ then $z' = \mu z$ and so $1 = z'_k = \mu z_k = \mu$. Thus $\mu = 1$ and $z' = z$. □

Appendix D

Cotangent Bundles

The simplest symplectic manifolds beyond $(\mathbb{R}^{2n}, \sum_{i=1}^{n} dp_i \wedge dq^i)$ are the *cotangent bundles* of manifolds. We start by defining the symplectic structure on cotangent bundles, and use it to give a formulation of Noether's famous theorem relating symmetries and conserved quantities. Next, we introduce a class of Hamiltonians on cotangent bundles which generate (co)geodesic flows. This is all intended to illustrate the power and scope of the Hamiltonian formalism, but also serves as a source of examples in Chapter 4 (Examples 4.13 and 4.29). Finally, we give a geometric interpretation of the Hamilton–Jacobi equation.

Note: In this appendix, we use up-indices for components of vectors, down-indices for components of covectors. This is because of the appearance of metric tensors and Christoffel symbols in the section on cogeodesic flow.

D.1 Cotangent Bundles

Let Q be a manifold and $q \in Q$ be a point. Recall that a covector[1] η at q is a linear map $T_q Q \to \mathbb{R}$. We write $T_q^* Q$ for the space of covectors at q and $\pi : T^* Q \to Q$ for the *cotangent bundle* of covectors over Q. Recall that a 1-form is a section of the cotangent bundle, that is, a covector at each point.

Definition D.1 The cotangent bundle $T^* Q$ carries a *canonical 1-form* λ, defined as follows. Let $q \in Q$ and $\eta \in T_q^* Q$. If $v \in T_\eta T^* Q$ is a tangent vector to the cotangent bundle at the point η then $\lambda(v) := \eta(\pi_*(v))$.

Remark D.2 If we pick local coordinates $(q^1, \ldots, q^n)$ on Q and use the coordinates $p = \sum p_i \, dq^i$ for $p \in T_q^* Q$ then $\lambda = \sum p_i \, dq^i$.

Definition D.3 We call the 2-form $\omega = d\lambda$ the *canonical symplectic structure*

---

[1] I like to use η (eta) because η *eats* a vector and outputs a number.

on T^*Q; in local coordinates this is $\sum_{i=1}^n dp_i \wedge dq^i$, which makes it clear why it is symplectic.

The coordinates p_i, q^i are called *canonical coordinates* on T^*Q. They are canonical in the sense that, once the local coordinates $q^1, \ldots, q^n$ are chosen on Q, we get a basis $dq^1, \ldots, dq^n$ for the fibres T_q^*Q, and so we get fibre coordinates p_i for free.

Remark D.4 (Exercise D.20) Pick local coordinates q^i on a patch in Q and consider the Hamiltonian system $(q^1 \circ \pi, \ldots, q^n \circ \pi)$ on the π-preimage of this patch. Show that the canonical coordinates p_i are *minus* the Liouville coordinates associated with the global Lagrangian section given by the zero-section. Does the zero-section inherit an integral affine structure?

The fact that the p_i are canonical in the preceding sense has the consequence that changing coordinates on Q induces a *symplectic* change of canonical co-ordinates on T^*Q. Namely:

Lemma D.5 *If* $\psi : Q \to Q$ *is a diffeomorphism then*

$$\psi_* = (\psi^{-1})^* : T^*Q \to T^*Q, \quad \psi_*(\eta) = \eta \circ (d\psi^{-1}) \in T_{\psi(q)}^*Q \text{ for } \eta \in T_q^*Q$$

is a symplectomorphism of the cotangent bundle.

Proof This is immediate from the fact that the canonical 1-form (and hence ω) are defined without reference to coordinates, but you can see it explicitly as follows. Suppose $\psi : Q \to Q$ is a diffeomorphism (change of coordinates); we will write the new coordinates as $\psi^1(q), \ldots, \psi^n(q)$. The basis $d\psi^i$ is given by $d\psi^i = \sum \frac{\partial \psi^i}{\partial q^j} dq^j$. Let us write $\Psi := d\psi$ for the matrix $\frac{\partial \psi^i}{\partial q^j}$. The new fibre coordinates p_i' are chosen so that $\sum p_i' d\psi^i = \sum p_i dq^i$, so $p_i' = p_j(\Psi^{-1})_i^j$ (here, we think of p as a row vector so the matrix Ψ^{-1} multiplies it on the right). The change of coordinates $(p_i, q^j) \mapsto (p_i', \psi^j)$ is symplectic because

$$\sum_i dp_i' \wedge d\psi^i = \sum_{i,j,k} dp_j(\Psi^{-1})_i^j \wedge \Psi_k^i dq^k = \sum_j dp_j \wedge dq^j. \qquad \square$$

Remark D.6 (Exercise D.21) The observant reader will detect parallels, but also notice subtle differences, between Lemma 2.25 and Lemma D.5. Contemplate these parallels and differences, and then turn to Exercise D.21 and its solution to read more.

D.2 Cogeodesic Flow

Suppose that g is a metric on Q. Write $|\eta|$ for the length (with respect to g) of a covector $\eta \in T_q^*Q$.

Definition D.7 Consider the function $H : T^*Q \to \mathbb{R}$ defined by $H(\eta) = \frac{1}{2}|\eta|^2$. The Hamiltonian flow generated by H is called the *cogeodesic flow*.

Lemma D.8 *If $(p(t), q(t))$ is a flowline of the cogeodesic flow then $q(t)$ is a geodesic on Q for the metric g and $p(t)$ is g-dual to $\dot{q}(t)$, that is $p(t)(w) = g(\dot{q}(t), w)$ for all $w \in T_{q(t)}Q$.*

Proof In local coordinates q^i, let g_{ij} be the metric (symmetric in i and j) and g^{ij} be its inverse (i.e. $\sum_j g^{ij}g_{jk} = \delta_k^i$), so that if $\eta = \sum p_i\, dq^i$ then $H = \frac{1}{2}|\eta|^2 = \frac{1}{2}\sum_{i,j} g^{ij}p_i p_j$. Thus,

$$\dot{q}^k = \frac{\partial H}{\partial p_k} = \sum_j g^{kj}p_j \qquad \dot{p}_k = \frac{\partial H}{\partial q^k} = \frac{1}{2}\sum_{i,j}\frac{\partial g^{ij}}{\partial q^k}p_i p_j.$$

The first equation tells us that $p_j = \sum_k g_{jk}\dot{q}^k$ as desired. It remains to show that $q(t)$ is a geodesic. Substituting $p_k = \sum_j g_{jk}\dot{q}^j$ into the second equation gives

$$\dot{p}_k = \sum_{k,\ell}\frac{\partial g_{jk}}{\partial q^\ell}\dot{q}^j\dot{q}^\ell + \sum_j g_{jk}\ddot{q}^j = -\frac{1}{2}\sum_{i,j,\ell,m}\frac{\partial g^{ij}}{\partial q^k}g_{i\ell}g_{jm}\dot{q}^\ell\dot{q}^m. \qquad (\mathrm{D}.1)$$

By differentiating $\sum_j g^{ij}g_{jm} = \delta_m^i$ we get $\sum_{i,j}\frac{\partial g^{ij}}{\partial q^k}g_{i\ell}g_{jm} = -\frac{\partial g_{\ell m}}{\partial q^k}$. Rearranging Equation (D.1) gives

$$\ddot{q}^j = \sum_{k,\ell,m} g^{jk}\left(\frac{1}{2}\frac{\partial g_{\ell m}}{\partial q^k} - \frac{\partial g_{mk}}{\partial q^\ell}\right)\dot{q}^\ell\dot{q}^m.$$

Because we are summing over ℓ, m and $\dot{q}^\ell\dot{q}^m$ is symmetric in ℓ, m, we can rewrite this as

$$\ddot{q}^j = \frac{1}{2}\sum_{k,\ell,m} g^{jk}\left(\frac{\partial g_{\ell m}}{\partial q^k} - \frac{\partial g_{mk}}{\partial q^\ell} - \frac{\partial g_{\ell k}}{\partial q^m}\right)\dot{q}^\ell\dot{q}^m = -\sum_{\ell,m}\Gamma_{\ell m}^j\dot{q}^\ell\dot{q}^m,$$

where $\Gamma_{\ell m}^j$ are the Christoffel symbols. This is precisely the geodesic equation for the path $q(t)$. $\square$

Remark D.9 Note that $q(t)$ moves with speed $|\dot{q}(t)| = |p(t)|$. Since $H = \frac{1}{2}|p|^2$ is conserved along the Hamiltonian flow, this speed is constant. In other words, the geodesic is parametrised proportionally to arc-length, where the constant of proportionality depends on the level set of H.

D.3 Noether's Theorem

The Hamiltonian formalism assigns a Hamiltonian flow ϕ_t^H to a function H on a symplectic manifold; the function H is conserved along the flow in the sense that $H(\phi_t^H(x)) = H(x)$ for all t. This is responsible for Noether's famous correspondence between symmetries and conserved quantities. We reiterate here that H does not need to correspond to 'energy': it can be any function.

Example D.10 (Translation)   Consider the symplectic manifold $\mathbb{R}^2$ with coordinates (p, q) and symplectic form $dp \wedge dq$. This is the phase space of a particle on a line, where q denotes the position of the particle and p its momentum. The Hamiltonian vector field associated with the function p is

$$\dot{p} = 0, \quad \dot{q} = 1,$$

which generates the flow $\phi_t^p(p, q) = (p, q + t)$. This is a translation in the q-direction.

More generally, if ξ is a vector field on Q and $\psi_t : Q \to Q$ is its flow, Noether's theorem gives an explicit Hamiltonian on T^*Q generating the 1-parameter family of symplectomorphisms $(\psi_t)_* : T^*Q \to T^*Q$ defined in Lemma D.5:

Theorem D.11 (Noether's theorem)   *Let ξ be a vector field on Q and let ψ_t be its flow. Define $H_\xi : T^*Q \to \mathbb{R}$ by $H_\xi(\eta) = \eta(\xi(q))$ for $\eta \in T_q^*Q$. If $\phi_t^{H_\xi}$ is the Hamiltonian flow of H_ξ then $\phi_t^{H_\xi} = (\psi_t)_*$, where $(\psi_t)_*$ is the action of ψ_t on T^*Q defined in Lemma D.5.*

Remark D.12   Let $\mathfrak{diff}(Q)$ denote the space of vector fields on Q (the Lie algebra of $\mathrm{Diff}(Q)$) and define the map $\mu : T^*Q \to (\mathfrak{diff}(Q))^*$ by $\mu(\eta)(\xi) = H_\xi(\eta)$. In the language of Chapter 3, Theorem D.11 can be phrased by saying that μ is a moment map for the Hamiltonian $\mathrm{Diff}(Q)$ action on T^*Q.

Proof   Write $\eta = \sum p_i \, dq^i$. We have $H_\xi(p, q) = \eta(\xi(q)) = \sum p_i \xi^i$. Both $\phi_t^{H_\xi}$ and $(\psi_t)_*$ are generated by vector fields. It suffices to check that these vector fields coincide.

By definition, $\phi_t^{H_\xi}$ is generated by the Hamiltonian vector field V_{H_ξ}. We claim that

$$V_{H_\xi} := -\sum_{i,j} p_i \frac{\partial \xi^i}{\partial q^j} \frac{\partial}{\partial p_j} + \sum_j \xi^j \frac{\partial}{\partial q^j}.$$

To see this, observe that

$$\iota_{V_{H_\xi}} \sum dp_k \wedge dq^k = \sum dp_k(V_{H_\xi})\,dq^k - \sum dq^k(V_{H_\xi})\,dp_k$$

$$= -\sum p_i \frac{\partial \xi^i}{\partial q^k}\,dq^k - \sum \xi^k\,dp_k = -dH_\xi.$$

Now let us calculate the infinitesimal action of ξ on T^*Q. Suppose that ψ_t is the flow of ξ; in coordinates, $(\psi_t)_*(p_i, q^i) = \left(\sum_j p_j \frac{\partial(\psi_t^{-1})^j}{\partial q^i}, \psi_t^i(q)\right)$. The infinitesimal action of ξ on T^*Q is given by

$$\left.\frac{d}{dt}\right|_{t=0} (\psi_t)_*(p_i, q^i) = \left.\frac{d}{dt}\right|_{t=0} \left(\sum_j p_j \frac{\partial(\psi_t^{-1})^j}{\partial q^i}, \psi_t^i(q)\right)$$

$$= \left(-\sum p_j \frac{\partial \xi^j}{\partial q^i}, \xi^i\right).$$

This is just another way of writing V_{H_ξ}, so the theorem is proved. In getting to the final line, we used the fact that $\left.\frac{d}{dt}\right|_{t=0} \psi_t^{-1} = -\xi$, that is, the inverse of ψ_t is obtained by flowing backwards along ξ for time t. $\qquad\square$

Example D.13 (Angular momentum, Exercise D.22) Suppose $Q = \mathbb{R}^3$ with coordinates q^1, q^2, q^3, and consider the 1-parameter family of diffeomorphisms

$$\psi_t(q^1, q^2, q^3) = (q^1 \cos t - q^2 \sin t, q^1 \sin t + q^2 \cos t, q^3)$$

given by rotating around the q^3-axis. Find the Hamiltonian on T^*Q which generates $(\psi_t)_*$.

Remark D.14 For those who have encountered Noether's theorem in the context of classical field theory, the field theory version is proved in the same way, where we take Q to be the space of fields and T^*Q to be the phase space.

D.4 Lagrangian Submanifolds and the Hamilton–Jacobi Equation

There are some easy examples of Lagrangian submanifolds in cotangent bundles.

Example D.15 The *zero-section* is the submanifold which intersects every cotangent fibre at the zero covector. This is Lagrangian: in local canonical coordinates (p, q), with $\omega = \sum_i dp_i \wedge dq^i$, it is given by $p = 0$. Dually, the cotangent fibres $q = $ const are also Lagrangian submanifolds.

Example D.16 Suppose $H: T^*Q \to \mathbb{R}$ is the Hamiltonian from Definition D.7 generating the cogeodesic flow on T^*Q for some metric. What are the geodesics connecting $x \in Q$ to $y \in Q$ in time t? They are in bijection with the intersection points between the Lagrangian submanifolds $\phi_t^H(T_x^*Q)$ and T_y^*Q. In this way, Lagrangian submanifolds can be used to impose initial/terminal conditions on geodesics or other Hamiltonian systems. The utility of this stems from the fact there is a variational interpretation for Hamilton's equations with Lagrangian boundary conditions (the Hamiltonian trajectories are critical points for the *action functional*). This is the point of departure for applications of Floer theory to symplectic geometry.

Recall that a *section* of the cotangent bundle is a map $\eta: Q \to T^*Q$ such that $\eta(q) \in T_q^*Q$. This is the same thing as a 1-form on Q. We define the *graph* of a 1-form η to be the image of the corresponding section $\eta(Q) = \{(q, \eta(q) \in T^*Q : q \in Q\} \subseteq T^*Q$. This is a submanifold diffeomorphic to Q.

Lemma D.17 *The graph of a 1-form η is Lagrangian if and only if $d\eta = 0$, that is, η is closed.*

Proof Let q^i be local coordinates on Q. The tangent space to $\eta(Q)$ at a point living over this coordinate patch is spanned by the vectors

$$\eta_*(\partial_{q^i}) = \frac{\partial \eta_j}{\partial q^i} \partial_{p_j} + \partial_{q^i}.$$

We have

$$\left(\sum_m dp_m \wedge dq^m \right) (\eta_* \partial_{q^k}, \eta_* \partial_{q^\ell}) = \frac{\partial \eta_\ell}{\partial q^k} - \frac{\partial \eta_k}{\partial q^\ell}.$$

This is the $dq^k \wedge dq^\ell$-component of $d\eta$. (Compare this with the proof of Theorem 1.42.) □

Note that if $L \subseteq T^*Q$ is transverse to the cotangent fibres near some point $x \in L$ then, locally near x, L is the graph of some section (by the inverse function theorem). Moreover, locally, any closed 1-form η admits an antiderivative, that is, a function S such that $\eta = dS$. We often call such a function S a (local) *generating function* for $\eta(Q)$. So to describe a Lagrangian submanifold of T^*Q, away from points where it is tangent to cotangent fibres, it is sufficient to give a collection of local generating functions.

If a Lagrangian submanifold is allowed to evolve under a Hamiltonian flow, then (up to a time-dependent constant shift) its local generating functions evolve according to a differential equation called the *Hamilton–Jacobi equation*. We state this in its simplest form for Lagrangians that admit a global generating function, namely Lagrangians which are the graph of an exact 1-form.

Theorem D.18 (Hamilton–Jacobi equation) *Let $L = \text{graph}(dS) \subseteq T^*Q$ be a Lagrangian submanifold which is the graph of an exact 1-form dS. Let $H_t : T^*Q \to \mathbb{R}$ be a time-dependent Hamiltonian. If S_t is a solution to the Hamilton–Jacobi equation*

$$\frac{\partial S_t}{\partial t} = -H_t\left(\frac{\partial S_t}{\partial q}, q\right), \qquad S_0 = S, \qquad (D.2)$$

then

$$\text{graph}(dS_t) = \phi_t^{H_t}(\text{graph}(dS)).$$

Conversely, if $\phi_t^{H_t}(\text{graph}(dS)) = \text{graph}(dF_t)$ then $F_t = S_t + c(t)$ where S_t solves Equation (D.2) and $c(t)$ is a time-dependent constant.

Proof Let $dS\colon Q \to T^*Q$ be the section corresponding to the 1-form dS and let $i_t\colon Q \to T^*Q$ be the Lagrangian inclusion of $\phi_t^{H_t}(dS(Q))$ defined by $i_t(q) = \phi_t^{H_t}(dS(q))$. Pick local canonical coordinates (p, q) and write $i_{t+\epsilon}(q) = i_t(q) + \epsilon v_t(q)$ for some vector field v_t along $i_t(Q)$. To first order in ϵ, $v_t = V_{H_t}$, that is:

$$v_t = \left(-\frac{\partial H}{\partial q} + \mathrm{o}(\epsilon), \frac{\partial H}{\partial p} + \mathrm{o}(\epsilon)\right)$$

where we write $\mathrm{o}(\epsilon)$ for any terms such that $\lim_{\epsilon \to 0} |\mathrm{o}(\epsilon)| = 0$.

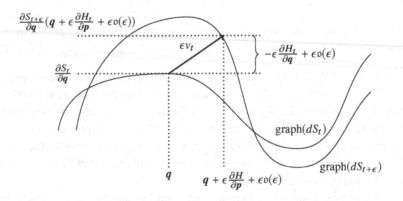

Figure D.1 The graphs of dS_t and $dS_{t+\epsilon}$ differ by $\epsilon v_t = \epsilon(V_{H_t} + \mathrm{o}(\epsilon))$.

We have

$$\frac{\partial S_{t+\epsilon}}{\partial q^i}\left(q + \epsilon\frac{\partial H}{\partial p} + o(\epsilon)\right) = \frac{\partial S_t}{\partial q^i}(q) - \epsilon\frac{\partial H_t}{\partial q^i}\left(\frac{\partial S_t}{\partial q}(q), q\right) + o(\epsilon)$$

and, by Taylor expanding, we also have

$$\frac{\partial S_{t+\epsilon}}{\partial q^i}\left(q + \epsilon\frac{\partial H}{\partial p} + o(\epsilon)\right) = \frac{\partial S_t}{\partial q^i}(q) + \epsilon\left(\frac{\partial^2 S_t}{\partial t\partial q^i}(q)+\right.$$

$$\left. + \sum_j \frac{\partial^2 S_t}{\partial q^i\partial q^j}(q)\frac{\partial H_t}{\partial p_j}\right) + o(\epsilon).$$

Comparing terms of order ϵ, we get

$$\frac{\partial^2 S_t}{\partial q^i\partial t} + \frac{\partial H_t}{\partial q^i} + \sum_j \frac{\partial S_t}{\partial q^j}\frac{\partial H_t}{\partial p_j} = 0,$$

where the derivatives of H_t are evaluated at $(\partial S_t/\partial q, q)$. In particular, this means that $\frac{\partial H_t}{\partial q^i} + \sum_j \frac{\partial S_t}{\partial q^j}\frac{\partial H_t}{\partial p_j} = \frac{\partial}{\partial q^i}((dS_t)^* H_t)$, so the equation is telling us that

$$\frac{\partial S_t}{\partial t} + (dS_t)^* H_t$$

is constant on Q, say equal to $C(t)$. This means that S_t satisfies the equation

$$\frac{\partial S_t}{\partial t} = -H_t\left(\frac{\partial S_t}{\partial q}, q\right) + C(t),$$

which means that $S_t - c(t)$ satisfies the Hamilton–Jacobi equation provided $\dot{c}(t) = C(t)$. □

Remark D.19 The proof is purely local, and therefore also works when the generating function is local, but it is trickier to state in that case because the domain of the local generating function changes.

D.5 Solutions to Inline Exercises

Exercise D.20 (Remark D.4) *Pick local coordinates q^i on a patch in Q and consider the Hamiltonian system $(q^1 \circ \pi, \ldots, q^n \circ \pi)$ on the π-preimage of this patch. Show that the canonical coordinates p_i are minus the Liouville coordinates associated with the global Lagrangian section given by the zero-section. Does the zero-section inherit an integral affine structure?*

Solution In local coordinates (p, q), $p = (p_1, \ldots, p_n)$ and $q = (q^1, \ldots, q^n)$, with $\omega = \sum dp_i \wedge dq^i$, the Hamiltonian flow of q^i is translation in the $-p_i$

direction, and the zero-section is given by $\sigma(q) = (q, 0)$. We have $\phi_t^q(q, 0) = (q, -t)$, which shows that the $-p_i$ are Liouville coordinates. Since the fibres of π are $\mathbb{R}^n$, there are no periodic orbits, so the period lattice is 0 in each fibre. Therefore, there is no natural integral affine structure: that construction would need the period lattice to have full rank. □

Exercise D.21 (Remark D.6) *Explain the parallels and differences between Lemmas 2.25 and D.5.*

Solution By Exercise D.20, we can think of $\pi\colon T^*Q \to Q$ as a Hamiltonian system. In both lemmas, we assume the existence of a diffeomorphism between the images of our Hamiltonian systems: for Lemma 2.25 we have $\phi\colon F(X) \to G(X)$ and for Lemma D.5 we have $\psi\colon Q \to Q$. In both cases, we obtain a symplectomorphism between the total spaces: respectively $\Phi\colon X \to Y$ and $(\psi)_*\colon T^*Q \to T^*Q$. Moreover, these symplectomorphisms are given by the same formula in Liouville coordinates.

The difference is that ϕ is required to be an integral affine transformation, whereas ψ can be any diffeomorphism. This is because the Hamiltonian systems F and G have period lattices of rank n, and the derivative of ϕ is required to preserve these period lattices, which tells us that ϕ is integral affine. The period lattice for π is trivial, so there is no constraint on $d\psi$. □

Exercise D.22 (Angular momentum, Example D.13) *Suppose $Q = \mathbb{R}^3$ with coordinates q^1, q^2, q^3, and consider the 1-parameter family of diffeomorphisms*

$$\psi_t(q^1, q^2, q^3) = (q^1 \cos t - q^2 \sin t, q^1 \sin t + q^2 \cos t, q^3)$$

given by rotating around the q^3-axis. Find the Hamiltonian on T^*Q which generates $(\psi_t)_*$.

Solution The flow ψ_t is generated by the vector field $\xi = (-q_2, q_1, 0)$, so by Theorem D.11, the induced symplectomorphism on T^*Q is generated by the Hamiltonian

$$H_\xi(p, q) = p_2 q_1 - p_1 q_2.$$

This is the usual formula for the component of angular momentum around the q^3-axis. □

Appendix E

Moser's Argument

At various points in the book, we have appealed to *the Moser argument*. This is a famous and extremely useful trick, first introduced by Moser [82]. We include a proof here for completeness. When we say a 'family of k-forms', we mean a k-form whose coefficients (with respect to any local coordinate system) depend continuously-differentiably on a parameter t.

Theorem E.1 (Moser's argument)   *Suppose that X is a manifold and ω_t is a family of symplectic forms. If $d\omega_t/dt = d\sigma_t$ for some family of compactly-supported 1-forms σ_t then there is a family of diffeomorphisms ϕ_t with $\phi_0 = $ id and $\phi_t^* \omega_t = \omega_0$.*

Proof   Let V_t be the vector field ω_t-dual to $-\sigma_t$, that is $\iota_{V_t}\omega_t = -\sigma_t$. This is a compactly-supported vector field, so we can define the flow along V_t. The flow is a 1-parameter family of diffeomorphisms ϕ_t satisfying $\phi_0 = $ id and $\frac{d\phi_t(x)}{dt} = V_t(\phi_t(x))$. We will differentiate $\phi_t^* \omega_t$ with respect to t and show that the result is zero. This will imply that $\phi_t^* \omega_t$ is independent of t, and hence equal to ω_0:

$$\frac{d}{dt}\phi_t^* \omega_t = \phi_t^* \left(\mathcal{L}_{V_t} \omega_t \right) + \phi_t^* \frac{d\omega_t}{dt}$$
$$= \phi_t^* (d\iota_{V_t}\omega_t) - \phi_t^* d\sigma_t$$
$$= \phi_t^* (d\sigma_t - d\sigma_t) = 0,$$

where we used Cartan's formula $\mathcal{L}_{V_t}\omega_t = d\iota_{V_t}\omega_t + \iota_{V_t}d\omega_t$ and the fact that $d\omega_t = 0$.   □

Appendix F

Toric Varieties Revisited

In this appendix, we will construct the toric variety associated to a convex rational polytope using only algebraic geometry (no symplectic cuts). Since most expositions of toric geometry (for example, Danilov [23] or Fulton [40]) start from the dual (fan) picture, and we are aiming to give alternative viewpoints wherever possible, we will confine ourselves to work only with the moment polytope. Throughout this appendix we will make use of homogeneous coordinates; see Appendix C for a rapid overview.

F.1 Construction

Let $\Delta \subseteq \mathbb{R}^n$ be a compact Delzant polytope. We will focus on the special case where the vertices of Δ have integer coordinates and explain how to construct the manifold X_Δ whose existence is guaranteed by Delzant's existence theorem, Theorem 3.7(2).

Theorem F.1 *Suppose that $p_1, \ldots, p_N \in \mathbb{Z}^n$ are the integer lattice points contained in a Delzant polytope Δ, and write $p_i = (p_{i1}, \ldots, p_{in})$. Let z^{p_i} be the monomial $z_1^{p_{i1}} z_2^{p_{i2}} \cdots z_n^{p_{in}}$. Consider the map*

$$F_\Delta \colon (\mathbb{C}^*)^n \to \mathbb{CP}^{N-1}, \qquad F_\Delta(z) = [z^{p_1} : \cdots : z^{p_N}].$$

Let X_Δ be the Zariski-closure of the image of F_Δ. Then X_Δ is a smooth projective variety. Let $P \colon \mathbb{R}^N \to \mathbb{R}^n$ be the linear projection given by right-multiplication with the matrix

$$\begin{pmatrix} p_{11} & p_{21} & \cdots & p_{n1} \\ p_{12} & \ddots & & p_{n2} \\ \vdots & & \ddots & \vdots \\ p_{1N} & \cdots & \cdots & p_{Nn} \end{pmatrix}.$$

176

If $\mu\colon \mathbb{CP}^N \to \mathbb{R}^N$ is the moment map for the standard T^N-action then

$$\mu|_{X_\Delta} \cdot P\colon X_\Delta \to \mathbb{R}^n$$

is the moment map for a T^n-action on X_Δ whose moment image is Δ.

Definition F.2 The projective variety X_Δ is called the *projective toric variety* associated to the polytope Δ.

Before proving this theorem, we will work out some examples.

F.2 Examples

Example F.3 Suppose Δ is the square with vertices $(0,0)$, $(1,0)$, $(0,1)$, and $(1,1)$. Since this is a square, Delzant's uniqueness theorem tells us that X_Δ will be $S^2 \times S^2$. We will confirm that this is the output of Theorem F.1.

Theorem F.1 tells us to consider the map

$$F_\Delta\colon (\mathbb{C}^*)^2 \to \mathbb{CP}^3, \qquad F_\Delta(z_1, z_2) = [1 : z_1 : z_2 : z_1 z_2].$$

If $[x_1 : x_2 : x_3 : x_4]$ are our homogeneous coordinates on $\mathbb{CP}^3$ then we see that the image of F_Δ is contained (as a Zariski-dense subset) in the subvariety $V = \{x_1 x_4 = x_2 x_3\}$. This subvariety is a smooth quadric surface and it is the Zariski-closure of the image of F_Δ. Note that F_Δ is the restriction of the Segre embedding

$$\mathbb{CP}^1 \times \mathbb{CP}^1 \to \mathbb{CP}^3, \qquad ([a : b], [c : d]) \mapsto [ac : bc : ad : bd]$$

to the affine chart $a = c = 1$, and V is the image of the Segre embedding. Since $\mathbb{CP}^1 \cong S^2$, this confirms that $X_\Delta = S^2 \times S^2$.

The matrix P is

$$\begin{pmatrix} 0 & 0 \\ 1 & 0 \\ 0 & 1 \\ 1 & 1 \end{pmatrix}$$

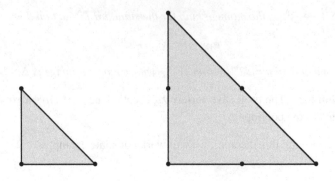

Figure F.1 The polygons Δ and 2Δ for Examples F.4 and F.5.

and the moment map μ is

$$\left(\frac{1}{2}\frac{|x_1|^2}{|x|^2},\ \frac{1}{2}\frac{|x_2|^2}{|x|^2},\ \frac{1}{2}\frac{|x_3|^2}{|x|^2},\ \frac{1}{2}\frac{|x_4|^2}{|x|^2}\right),$$

so

$$\mu(x)\cdot P = \left(\frac{1}{2}\frac{|x_2|^2+|x_4|^2}{|x|^2},\ \frac{1}{2}\frac{|x_3|^2+|x_4|^2}{|x|^2}\right).$$

Precomposing with the Segre embedding to get a function on $X_\Delta = \mathbb{CP}^1 \times \mathbb{CP}^1$, we get

$$\left(\frac{1}{2}\frac{|b|^2}{|a|^2+|b|^2},\ \frac{1}{2}\frac{|d|^2}{|c|^2+|d|^2}\right).$$

The first (respectively second) component is the Hamiltonian generating the standard circle action on the first (respectively second) factor $\mathbb{CP}^1$ (see Example 3.17).

The moment image is the convex hull of the moment images of the fixed points. The fixed points are $([1:0],[1:0])$, $([1:0],[0:1])$, $([0:1],[1:0])$, and $([0:1],[0:1])$, whose images are $(0,0)$, $(1,0)$, $(0,1)$ and $(1,1)$ respectively. Therefore, the moment image of X_Δ is Δ.

Example F.4 If Δ is the triangle with vertices $(0,0)$, $(1,0)$ and $(0,1)$ then $F_\Delta\colon (\mathbb{C}^*)^2 \to \mathbb{CP}^2$ is the map $F_\Delta(z_1,z_2) = [1:z_1:z_2]$. The image of F_Δ is dense in $\mathbb{CP}^2$, so $X_\Delta = \mathbb{CP}^2$. By Lemmas 3.16 and 3.20, the preimage of an edge is a symplectic sphere with area 2π and self-intersection 1; this is a line in $\mathbb{CP}^2$.

If we rescale both the Fubini–Study form on $\mathbb{CP}^2$ and the moment map for the torus action by a factor of 2 then we get a moment map whose image is the triangle 2Δ with vertices $(0,0)$, $(2,0)$ and $(0,2)$. Delzant's uniqueness theorem tells us that $X_{2\Delta} \cong (\mathbb{CP}^2, 2\omega_{FS})$. But the isomorphism is not obvious from the construction.

Example F.5 If 2Δ is the triangle with vertices $(0,0)$, $(2,0)$ and $(0,2)$ then there are now six integer points in 2Δ, and we get

$$F_\Delta: (\mathbb{C}^*)^2 \to \mathbb{CP}^5, \qquad F_\Delta(z_1, z_2) = [1 : z_1 : z_1^2 : z_2 : z_1 z_2 : z_2^2].$$

The map F_Δ factors through the *quadratic Veronese embedding*

$$\mathcal{V}: \mathbb{CP}^2 \to \mathbb{CP}^5, \qquad \mathcal{V}([a : b : c]) = [a^2 : ab : b^2 : ac : bc : c^2]$$

by taking $a = 1$, $b = z_1$, $c = z_2$, and the image of F_Δ is dense inside $\mathcal{V}(\mathbb{CP}^2)$. Thus $X_\Delta = \mathbb{CP}^2$. Again, the preimage of an edge is a line in $\mathbb{CP}^2$, but it has symplectic area 4π because the pullback of the Fubini–Study form along the quadratic Veronese embedding is symplectomorphic to *twice* the Fubini–Study form on $\mathbb{CP}^2$ (a hyperplane of $\mathbb{CP}^5$ intersects $\mathcal{V}(\mathbb{CP}^2)$ in a conic, not a line).

Rescaling the polytope by a factor of k always corresponds to reimbedding via a Veronese map of degree k.

Finally, let us try to apply the construction from Theorem F.1 when Δ is not Delzant. The corresponding toric variety will have singularities living over the non-Delzant points of Δ.

Example F.6 Let Δ be the non-Delzant polygon with vertices $(0,0)$, $(0,1)$ and $(2,1)$ from Figure 3.1. This additionally contains the integer point $(1,1)$. We therefore get

$$F_\Delta: (\mathbb{C}^*)^2 \to \mathbb{CP}^3, \qquad F_\Delta(z_1, z_2) = [1 : z_2 : z_1 z_2 : z_1^2 z_2].$$

In homogeneous coordinates $[x_1 : x_2 : x_3 : x_4]$ this satisfies the equation $x_2 x_4 = x_3^2$. This is a singular quadric surface with an ordinary double point at $[1 : 0 : 0 : 0]$. Under the moment map $\mu \cdot P$, this point projects to the origin, which is precisely the point where Δ fails to be Delzant.

By Lemma 3.20, the preimage of the horizontal edge is a symplectic sphere with square 2. In homogeneous coordinates, this is the conic $x_2 x_4 = x_3^2$ in the plane $x_1 = 0$.

Remark F.7 The ordinary double point is the cyclic quotient singularity $\frac{1}{2}(1, 1)$. The germ of our non-Delzant polygon near the origin agrees with the germ of the non-Delzant polygon from Example 3.21 with $n = 2$, $a = 1$. This is a general fact: one can read off the singularities of X_Δ from the non-Delzant points in Δ.

F.3 Proof of Theorem F.1

Consider the T^n-actions

$$e^{it} z = (e^{it_1} z_1, \ldots, e^{it_n} z_n),$$

$$e^{it} \star [Z_1 : \ldots : Z_N] = \left[e^{i(p_{11}t_1 + \cdots + p_{1n}t_n)} Z_1 : \cdots : e^{i(p_{N1}t_1 + \cdots + p_{Nn}t_n)} Z_N \right]$$

on $(\mathbb{C}^*)^n$ and $\mathbb{CP}^N$ respectively. The action denoted by $\star$ is generated by the Hamiltonian $\mu \cdot P$. The map F_Δ intertwines the actions in the sense that $F_\Delta(e^{it} z) = e^{it} \star F_\Delta(z)$; this means that $e^{it} \star$ preserves the image of F_Δ, and hence its Zariski-closure X_Δ.

It remains to show that X_Δ is smooth and that the moment image agrees with Δ. We will start by writing down equations for X_Δ. Let $[Z_1 : \cdots : Z_N]$ be homogeneous coordinates on $\mathbb{CP}^{N-1}$. Note that each coordinate Z_i corresponds to an integer lattice point $p_i \in \Delta$.

Lemma F.8 *Let $a_1, \ldots, a_N$ be integers. If the relation $\sum_i a_i p_i = 0$ holds then the equation*

$$\prod_{a_i \geq 0} Z_i^{a_i} = \prod_{a_i < 0} Z_i^{-a_i}$$

holds on X_Δ.

Proof This holds on the image of F_Δ because it translates to $\prod_i z^{\sum a_i p_i} = z^0 = 1$. It therefore holds on the Zariski-closure of the image of F_Δ, which is X_Δ by definition. □

Let V_Δ be the subvariety cut out by the equations coming from Lemma F.8. The lemma shows that $X_\Delta \subseteq V_\Delta$. We will show that V_Δ is a smooth variety containing the image of F_Δ as a Zariski-open set, which will show that $V_\Delta = X_\Delta$ (in particular, it will show that X_Δ is smooth).

Let $\Gamma = \{v \in \{1, \ldots, N\} : p_v$ is a vertex of $\Delta\}$. Note that every integer lattice point in Δ can be written as a linear combination $\sum_{j \in \Gamma} a_j p_j$ with $a_j \in \mathbb{Q}_{\geq 0}$ for all $j \in \Gamma$.

Corollary F.9 *For each $i \in \{1, \ldots, N\}$, write $p_i = \sum_{j \in \Gamma(\Delta)} a_j p_j$ with $a_j \in \mathbb{Q}_{\geq 0}$. Let $\Gamma_i = \{j \in \Gamma : a_j \neq 0\} \subseteq \Gamma$. The open set $V_\Delta \cap \{Z_i \neq 0\}$ is contained in the intersection*

$$V_\Delta \cap \bigcap_{j \in \Gamma_i} \{Z_j \neq 0\}.$$

In particular, V_Δ is covered by the open sets $V_\Delta \cap \{Z_j \neq 0\}$, $j \in \Gamma$.

Proof Let $b \in \mathbb{Z}_{>0}$ be such that $c_j := ba_j \in \mathbb{Z}_{\geq 0}$. The equation $Z_i^b = \prod_{j \in \Gamma_i} Z_j^{c_j}$ holds on V_Δ by Lemma F.8. If $Z_i \neq 0$ then this means $Z_j \neq 0$ for all $j \in \Gamma_i$. $\qquad\square$

Lemma F.10 (Exercise F.14) *Let A be an integer matrix with rows A_i. Consider the morphism $\tilde{A} \colon (\mathbb{C}^*)^n \to (\mathbb{C}^*)^n$ defined by*

$$\tilde{A}(z) = (z^{A_1}, \ldots, z^{A_n}).$$

If we write $w := \tilde{A}(z)$ then $w^q = z^{qA}$ for any integer row vector q. The morphism $\tilde{A}$ is invertible if and only if $A \in GL(n, \mathbb{Z})$.

Lemma F.11 *Suppose $T \colon \mathbb{R}^n \to \mathbb{R}^n$ is a map of the form $T(x) = xA + c$ for some $A \in GL(n, \mathbb{Z})$ and $c \in \mathbb{Z}^n$. Let $\tilde{A}^{-1} \colon (\mathbb{C}^*)^n \to (\mathbb{C}^*)^n$ be the morphism given by the matrix A^{-1} as in Lemma F.10. Then $F_{T(\Delta)} \circ \tilde{A}^{-1} = F_\Delta$.*

Proof First note that the constant term c has no effect on the image of F_Δ: it introduces an overall scale factor z^c into every homogeneous coordinate. We therefore assume without loss of generality that $c = 0$. Let $q_i = p_i A$ be the vertices of $T(\Delta)$. We have:

so
$$F_{T(\Delta)}(w) = [w^{q_1} : \cdots : w^{q_N}],$$

$$F_{T(\Delta)} \circ \tilde{A}^{-1}(z) = [z^{q_1 A^{-1}} : \cdots : z^{q_N A^{-1}}]$$
$$= [z^{p_1} : \cdots : z^{p_N}] = F_\Delta(z). \qquad\square$$

Remark F.12   Note that although the variety X_Δ is unchanged by T, the moment map is changed by T because the projection P from Theorem F.1 changes in such a way that the moment image is $T(\Delta)$.

Lemma F.13 *If $i \in \Gamma$ then $V_\Delta \cap \{z_i \neq 0\}$ is T^n-equivariantly biholomorphic to $\mathbb{C}^n$ with its standard torus action.*

Proof Because our polytope is Delzant, we can apply a transformation as in Lemma F.11 so that p_i is at the origin. By making a further transformation, we can assume that if $p_{j_1}, \ldots, p_{j_n}$ are the closest lattice points to p_i along the n edges meeting at p_i then these sit at the points $(1, 0, \ldots, 0), \ldots, (0, \ldots, 0, 1)$. Now any lattice point $p_k \in \Delta$ can be written as a nonnegative integer linear combination of these basis vectors, so $Z_k = \prod_{s=1}^n Z_{j_s}^{a_{j_s}}$ with $a_{j_s} \in \mathbb{Z}_{\geq 0}$. This means that on $Z_i \neq 0$ we can take $Z_i = 1$ and use $Z_{j_1}, \ldots, Z_{j_n}$ as global coordinates on $V_\Delta \cap \{Z_i \neq 0\}$. Since p_{j_s} is the sth basis vector, the torus action rotates Z_{j_s} by e^{it_s}, which shows that the biholomorphism we have chosen is equivariant with the standard torus action. $\qquad\square$

As a consequence, we see that V_Δ is smooth because we have covered V_Δ by smooth coordinate charts. We also see that $F_\Delta((\mathbb{C}^*)^n) \subseteq V_\Delta$ is Zariski-dense in V_Δ because it intersects each chart $V_\Delta \cap \{Z_i \neq 0\} \cong \mathbb{C}^n$ in the Zariski-dense subset $(\mathbb{C}^*)^n$. We deduce that $X_\Delta = V_\Delta$ and that X_Δ is smooth.

Finally, we need to check that the moment image of X_Δ is Δ. Recall from Theorem 3.7(1) that the moment image is the convex hull of the moment images of the fixed points, so it suffices to show that the fixed points are sent by the moment map to the vertices of Δ.

For each $i \in \Gamma$ (i.e. p_i is a vertex of Δ), let $e_i \in \mathbb{CP}^{N-1}$ be the point whose homogeneous coordinates are $Z_i = 1$ and $Z_j = 0$ if $j \neq i$. In the T^n-equivariant local chart $Z_i \neq 0$ from Lemma F.13, e_i is sent to the origin, which is a T^n-fixed point and the only T^n-fixed point in that chart. This shows that the T^n-fixed points in X_Δ are precisely the points e_i. We have $\mu(e_i) \cdot P = p_i$, so we deduce that the T^n-fixed points map under the moment map to the vertices of Δ, as required.

F.4 Solutions to Inline Exercises

Exercise F.14 (Lemma F.10)   *Let A be an integer matrix with rows A_i. Consider the morphism $\tilde{A}\colon (\mathbb{C}^*)^n \to (\mathbb{C}^*)^n$ defined by*

$$\tilde{A}(z) = (z^{A_1}, \ldots, z^{A_n}).$$

If we write $w := \tilde{A}(z)$ then show that $w^q = z^{qA}$ for any integer row vector q. Prove that the morphism $\tilde{A}$ is invertible if and only if $A \in GL(n, \mathbb{Z})$.

Solution   We have $w_i = z_1^{A_{i1}} \cdots z_n^{A_{in}}$, so

$$
\begin{aligned}
w^q &= (z_1^{A_{11}q_1} \cdots z_n^{A_{1n}q_1}) \cdots (z_1^{A_{n1}q_n} \cdots z_n^{A_{nn}q_n}) \\
&= z_1^{A_{11}q_1 + \cdots + A_{n1}q_n} \cdots z_n^{A_{1n}q_1 + \cdots + A_{nn}q_n} \\
&= z^{qA}.
\end{aligned}
$$

In particular, this shows that $\widetilde{AB} = \tilde{A}\tilde{B}$.

In particular, if $A \in GL(n, \mathbb{Z})$ then $\widetilde{A^{-1}}$ gives an inverse for $\tilde{A}$.

More generally, let S, T be invertible integer matrices such that $D := SAT$ is in Smith normal form. Since $\tilde{D} = \tilde{S}\tilde{A}\tilde{T}$, we see that $\tilde{D}$ is invertible if and only if $\tilde{A}$ is invertible. But $\tilde{D}(z_1, \ldots, z_n) = (z_1^{D_{11}}, \ldots, z_n^{D_{nn}})$ which is invertible only if $D_{ii} = \pm 1$ for all i, which holds only if $D \in GL(n, \mathbb{Z})$, which holds only if $A \in GL(n, \mathbb{Z})$. $\square$

Appendix G

Visible Contact Hypersurfaces and Reeb Flows

We have focused a lot on visible Lagrangian submanifolds, but one can also 'see' other sorts of submanifolds using Lagrangian torus fibrations. In this section, we discuss visible submanifolds of codimension 1 and *contact geometry*.

G.1 Hypersurfaces

Suppose we have a Lagrangian fibration $f : X \to B$. One easy way to produce real codimension 1 hypersurfaces in X is to take the preimage of a codimension 1 submanifold of B.

Example G.1 Let $X = \mathbb{C}^2$, $B = \mathbb{R}^2$ and

$$f : X \to \mathbb{R}^2, \quad f(z_1, z_2) = \left(\frac{1}{2}|z_1|^2, \frac{1}{2}|z_2|^2 \right)$$

be the moment map for the standard torus action. Let $a, b, c > 0$ be positive constants and consider the line $\ell_{a,b,c} \subseteq B$ defined by the equation $ax + by = c$ (see Figure G.1). The preimage in X is the ellipsoid[1]

$$Y_{a,b,c} := \{ a|z_1|^2 + b|z_2|^2 = 2c \}.$$

Definition G.2 A vector field Z on a symplectic manifold (X, ω) is a *Liouville* or *symplectically dilating* vector field if $\mathcal{L}_Z \omega = \omega$. A hypersurface $Y \subseteq X$ is said to be of *contact-type* if there is a Liouville vector field defined in a neighbourhood of Y which is everywhere transverse to Y. If $Y = \partial X$ then we say Y is *convex* or *concave* if Z points respectively out of or into X.

---

[1] Note that *ellipsoid* is also used to mean the compact region bounded by this hypersurface (sometimes called a *solid ellipsoid*).

Figure G.1 The preimage of the line segment is an ellipsoid.

Example G.3 Continuing Example G.3, let (p_1, p_2, q_1, q_2) be action-angle coordinates on $\mathbb{C}^2$ for the standard torus action (so $p_i = \frac{1}{2}|z_i|^2$ and q_i is the argument of z_i) and let Z be the vector field given in action-angle coordinates by $p_1 \frac{\partial}{\partial p_1} + p_2 \frac{\partial}{\partial p_2}$. This is a Liouville vector field:

$$\mathcal{L}_Z \omega = d\iota_Z \omega = d(p_1 dq_1 + p_2 dq_2) = \sum dp_i \wedge dq_i.$$

Moreover, $f_* Z$ is the radial vector field in (the positive quadrant of) $\mathbb{R}^2$ (see Figure G.2). Since $a, b, c > 0$, this is transverse to $\ell_{a,b,c}$ and hence Z is transverse to $Y_{a,b,c}$. Thus our ellipsoids are contact-type hypersurfaces.

Figure G.2 There is a Liouville vector field transverse to the ellipsoid which projects to the radial vector field in action-coordinates.

More generally, this proves:

Lemma G.4 *Let (X, ω) be a symplectic manifold and suppose that (p, q) are local action-angle coordinates on a chart $U \subseteq X$. The vector field $\sum_i p_i \frac{\partial}{\partial p_i}$ is a Liouville vector field on U. If $H \colon U \to \mathbb{R}$ is a function which depends only on p and c is a regular value then $H^{-1}(c)$ is a contact-type hypersurface if and only if $\sum_i p_i \frac{\partial H}{\partial p_i}$ is nowhere vanishing on $H^{-1}(c)$.*

Proof The calculation from Example G.3 shows that Z is Liouville. To understand when $H^{-1}(c)$ is contact-type, we see that Z is transverse to $H^{-1}(c)$ if and only if $dH(Z) \neq 0$ everywhere along $H^{-1}(c)$. We compute:

$$dH(Z) = \sum_i \frac{\partial H}{\partial p_i} dp_i \left(\sum_j p_j \frac{\partial}{\partial p_j} \right) = \sum p_i \frac{\partial H}{\partial p_i},$$

which proves the result. □

G.2 Contact Forms and Reeb Flows

Definition G.5 A 1-form α on a $(2n-1)$-dimensional manifold Y is called a *contact form* if $\alpha \wedge (d\alpha)^{n-1}$ is nowhere zero.

Lemma G.6 *Let* $i : Y \to X$ *be the inclusion map for a contact-type hypersurface with transverse Liouville field* Z. *The 1-form* $\alpha := i^* \iota_Z \omega$ *is a contact form on* Y.

Proof The $2n$-form ω^n is a nowhere-vanishing volume form on X, so $i^* \iota_Z(\omega^n)$ is a nowhere-vanishing volume form on Y, since Z is transverse to Y. We have

$$\iota_Z(\omega^n) = (\iota_Z\omega) \wedge (\omega^{n-1}) + \cdots + (\omega^{n-1}) \wedge (\iota_Z\omega) = n(\iota_Z\omega) \wedge (\omega^{n-1}).$$

Therefore

$$\frac{i^* \iota_Z \omega^n}{n} = i^* \left(\iota_Z\omega \wedge \omega^{n-1} \right).$$

Set $\alpha = i^* \iota_Z\omega$. We have $d\alpha = i^* d\iota_Z\omega = i^* \mathcal{L}_Z\omega = i^*\omega$, so $\alpha \wedge d\alpha^{n-1} = i^* \left(\iota_Z\omega \wedge \omega^{n-1} \right)$. As this is a nowhere-vanishing volume form on Y, this proves the result. □

Definition G.7 If α is a contact form on Y, we define the *Reeb vector field* R_α to be the unique vector field satisfying

$$\iota_{R_\alpha} d\alpha = 0, \quad \iota_{R_\alpha} \alpha = 1.$$

Remark G.8 The equation $\iota_{R_\alpha} d\alpha = 0$ says that the Reeb field points along the line field[2] $(TY)^\omega \subseteq TY$. The equation $\iota_{R_\alpha} \alpha = 1$ is simply a normalisation condition which picks out a specific vector $(TY)^\omega$.

Example G.9 (Exercise G.11) Let X be a symplectic manifold and $U \subseteq X$ be the domain of an action-angle chart with action-angle coordinates (p, q). Let $H : U \to \mathbb{R}$ be a function depending only on p and satisfying the condition $\sum_i p_i \frac{\partial H}{\partial p_i} \neq 0$ along a regular level set $H^{-1}(c)$. Let $\alpha = i^*(\sum_k p_k \, dq_k)$ be the contact form guaranteed by Lemma G.4. The Reeb vector field is given by

$$R_\alpha = \left(\sum_i \frac{\partial H}{\partial p_i} \frac{\partial}{\partial q_i} \right) \bigg/ \left(\sum_j p_j \frac{\partial H}{\partial p_j} \right).$$

---

[2] This is called the *characteristic line field*. Recall that ω denotes the symplectic orthogonal complement; see Definition A.3.

Example G.10 (Exercise G.12)   Continuing Example G.3, we take $H = ap_1 + bp_2$, which gives the contact-type ellipsoid $Y_{a,b,c} \subseteq \mathbb{C}^2$. By Example G.9, the Reeb field is

$$c^{-1} \left(a \frac{\partial}{\partial q_1} + b \frac{\partial}{\partial q_2} \right).$$

The dynamics of the flow along the Reeb field now depend on the constants a and b.

If the ratio b/a is irrational then the orbits of R_α are lines of irrational slope in the (q_1, q_2)-torus. The exception is when $p_1 = 0$ or $p_2 = 0$: here the action-angle coordinates are degenerate in the sense that the fibre is a circle parametrised by q_2 respectively q_1. These two circles are closed orbits of the Reeb field (see Figure G.3).

Figure G.3   If the slope is irrational, then the only closed Reeb orbits are the circles living over the points marked ∘.

If the ratio b/a is rational then all the Reeb orbits are closed. More precisely, if $a = \rho m$ and $b = \rho n$ for coprime integers m, n then the orbits away from $p_1 = 0$ and $p_2 = 0$ have period $2\pi c/\rho$. There are still two exceptional orbits at $p_1 = 0$ and $p_2 = 0$, with periods $2\pi c/n\rho$ and $2\pi c/m\rho$ respectively. In this case, we can take the symplectic quotient of $\mathbb{C}^2$ with respect to this Hamiltonian and obtain a symplectic sphere with two orbifold points. (Compare with Example 4.12, where we obtained weighted projective spaces in this way.) The quotient map $Y_{a,b,c} \to Y_{a,b,c}/S^1$ is an example of a *Seifert fibration*; for more about the topology of Seifert fibred manifolds, see [55, 83] or Seifert's appendix to [94].

G.3 Solutions to Inline Exercises

Exercise G.11 (Example G.9)   *Show that the Reeb vector field in Example G.9 is given by*

$$R_\alpha = \left(\sum_i \frac{\partial H}{\partial p_i} \frac{\partial}{\partial q_i} \right) \Big/ \left(\sum_j p_j \frac{\partial H}{\partial p_j} \right).$$

Solution We recall that our contact manifold is the level set $Y := H^{-1}(c)$ of a function $H(\boldsymbol{p}, \boldsymbol{q})$ which is independent of $\boldsymbol{q}$. Since R_α has no ∂_{p_j} components, we have $dH(R_\alpha) = 0$, so R_α is tangent to Y. Next, we have $d\alpha = i^* \sum dp_k \wedge dq_k$, where i is the inclusion of the level set Y. Therefore,

$$\iota_{R_\alpha} d\alpha = \left(\sum_k \frac{\partial H}{\partial p_k} dp_k \right) \Big/ \left(\sum_j p_j \frac{\partial H}{\partial p_j} \right) \propto dH,$$

and dH vanishes on Y. Finally, observe that

$$\iota_{R_\alpha} \alpha = \left(\sum_k p_k \, dq_k \right) \left(\left(\sum_i \frac{\partial H}{\partial p_i} \frac{\partial}{\partial q_i} \right) \Big/ \left(\sum_j p_j \frac{\partial H}{\partial p_j} \right) \right)$$

$$= \left(\sum_i p_i \frac{\partial H}{\partial p_i} \right) \Big/ \left(\sum_j p_j \frac{\partial H}{\partial p_j} \right) = 1.$$

$\square$

Exercise G.12 (Example G.10) *Show that in Example G.10, the Reeb field is*

$$c^{-1} \left(a \frac{\partial}{\partial q_1} + b \frac{\partial}{\partial q_2} \right).$$

Solution We have $\partial H / \partial p_1 = a$ and $\partial H / \partial p_2 = b$, so the Reeb field from Example G.9 becomes

$$\left(a \frac{\partial}{\partial q_1} + b \frac{\partial}{\partial q_2} \right) \Big/ (a p_1 + b p_2).$$

Since $H(p, q) = a p_1 + b p_2 = c$ along $Y_{a,b,c}$, this is $c^{-1} \left(a \frac{\partial}{\partial q_1} + b \frac{\partial}{\partial q_2} \right)$ as required.

$\square$

Appendix H

Tropical Lagrangian Submanifolds

While visible Lagrangians are associated with straight line segments or affine subspaces in the base of a Lagrangian torus fibration, *tropical Lagrangians* are associated with certain *piecewise* linear subsets (like trivalent graphs). More precisely, a tropical Lagrangian is an immersed Lagrangian whose image under the fibration is a small thickening of a *tropical subvariety* in the base of the fibration. We will focus on the case of tropical curves.

Mikhalkin [80] and Matessi [72, 73] have given constructions of tropical Lagrangians associated with tropical curves and tropical hypersurfaces respectively. We will focus on the four-dimensional case where these constructions coincide, and we will use Mikhalkin's conventions.

H.1 A Lagrangian Pair-Of-Pants

We consider $\mathbb{C}^* \times \mathbb{C}^*$ equipped with complex coordinates (z_1, z_2). We identify this with $\mathbb{R}^2 \times T^2$ with coordinates $(p_1, p_2) \in \mathbb{R}^2$ and $(q_1, q_2) \in T^2 = (\mathbb{R}/2\pi\mathbb{Z})^2$ using the identification

$$z_k = \exp(p_k + i q_k).$$

Theorem H.1 (Mikhalkin) *Let R_1, R_2, R_3 be three rays with rational slope in the p-plane emanating from the origin, and let v_1, v_2, v_3 be the primitive integer vectors pointing along these rays. Suppose that the* balancing condition

$$v_1 + v_2 + v_3 = 0 \tag{H.1}$$

holds and that any two of these vectors form a $\mathbb{Z}$-basis for the integer lattice.[1] Let L_1, L_2, L_3 be the visible Lagrangian half-cylinders living over R_1, R_2, R_3 and fix $\epsilon > 0$. Let $U := \{p : |p| > \epsilon\}$. There is an embedded Lagrangian

---

[1] These three vectors form what Conway [21] calls a *superbase*.

submanifold $L \subseteq \mathbb{R}^2 \times T^2$, *diffeomorphic to the pair-of-pants, such that* $U \cap L = U \cap (L_1 \cup L_2 \cup L_3)$.

Proof We will focus on the case $v_1 = (-1, 0)$, $v_2 = (0, -1)$, $v_3 = (1, 1)$. This implies the general case: if v_1, v_2 is obtained from this basis by an element of $GL(2, \mathbb{Z})$ then Lemma 2.25 gives us a fibred symplectomorphism living over this integral affine transformation of the p-plane, and we can apply this symplectomorphism to the Lagrangian obtained for $v_1 = (-1, 0)$, $v_2 = (0, -1)$.

The starting point of the construction is the following exercise.

Lemma H.2 (Exercise H.11)   *Consider the* hyper-Kähler twist

$$\mathbb{R}^2 \times T^2 \to \mathbb{R}^2 \times T^2, \quad (p_1, p_2, q_1, q_2) \mapsto (p_1, p_2, -q_2, q_1).$$

If $C \subseteq \mathbb{R}^2 \times T^2$ *is a complex curve with respect to the complex coordinates* $z_k = e^{p_k + iq_k}$ *then the image of* C *under the hyper-Kähler twist is Lagrangian for the symplectic form* $\sum dp_i \wedge dq_i$.

We will apply this lemma to the complex curve $C = \{z_2 = 1 + z_1\} \subseteq \mathbb{C}^* \times \mathbb{C}^*$. This is diffeomorphic to a pair-of-pants (3-punctured sphere); we can parameterise it as $z \mapsto (z, 1 + z)$ with $z \in \mathbb{C} \setminus \{0, -1\}$. Let $L \subseteq \mathbb{R}^2 \times T^2$ be the hyper-Kähler twist of C. In coordinates, this is given parametrically by

$$p_1(z) = \ln|z|, \qquad\qquad p_2(z) = \ln|1 + z|,$$
$$q_1(z) = -\arg(1 + z), \qquad q_2(z) = \arg(z).$$

We now write $z = re^{i\theta}$ and see what happens as $r \to 0$. We have $\lim_{r \to 0} p_2(z) = \ln(1) = 0$ and $\lim_{r \to 0} q_1(z) = -\arg(1) = 0 \mod 2\pi$. This means that near the puncture 0, our Lagrangian L is asymptotic to the cylinder

$$L_1 := \{(p_1, 0, 0, q_2) : p_1 < 0, q_2 \in [0, 2\pi]\},$$

which is a visible Lagrangian cylinder associated to the negative p_1-axis (note that $\lim_{r \to 0} p_1(z) = -\infty$). We write R_1 for the negative p_1-axis. A similar analysis near the punctures $z \to -1$ and $z \to \infty$ shows that L has asymptotes along the visible Lagrangian cylinders

$$L_2 := \{(0, p_2, q_1, 0) : p_2 < 0, q_1 \in [0, 2\pi]\},$$
$$L_3 := \{(p, p, q, -q) : p > 0, q \in [0, 2\pi]\}$$

associated to the rays R_2 and R_3 shown in Figure H.1.

Given $0 < k_1 < k_2$, define the compact regions $K_i = \{p_1^2 + p_2^2 \le k_i\}$, $i = 1, 2$, and let $U_i = (\mathbb{C}^* \times \mathbb{C}^*) \setminus K_i$ be their complements. We will modify L to obtain a Lagrangian pair-of-pants L' with

$$K_1 \cap L' = K_1 \cap L \text{ and } U_2 \cap L' = U_2 \cap (L_1 \cup L_2 \cup L_3).$$

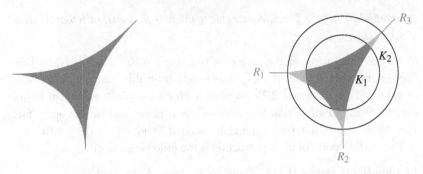

Figure H.1 (a) The projection of L to the $\boldsymbol{p}$-plane. (b) L is asymptotic to the visible Lagrangian cylinders L_1, L_2, L_3 which live over the three rays shown. We modify it outside the region K_1 so that it coincides with these visible Lagrangians outside K_2.

We will explain this modification for the puncture asymptotic to L_1; the other cases are similar. We can identify a neighbourhood of L_1 with a neighbourhood of the zero-section in T^*L_1. More precisely, we think of $p_1 = \ln r$ and $q_2 = \theta$ as coordinates on the cylinder and $-q_1$ and p_2 as dual momenta.[2]

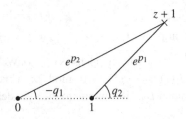

Figure H.2 Geometric picture behind formulae for q_1 and p_2.

From Figure H.2, you can extract equations for our section:

$$q_1 = -\arctan\left(\frac{e^{p_1}\sin q_2}{1 + e^{p_1}\cos q_2}\right), \qquad p_2 = \frac{1}{2}\ln(1 + 2e^{p_1}\cos q_2 + e^{2p_1}),$$

defined on the subset $p_1 < -\ln 2$. In other words, it is the graph of the 1-form

$$\beta := \frac{1}{2}\ln(1 + 2e^{p_1}\cos q_2 + e^{2p_1})dp_1 + \arctan\left(\frac{e^{p_1}\sin q_2}{1 + e^{p_1}\cos q_2}\right)dq_2.$$

This 1-form is closed (this is equivalent to the Lagrangian condition, but you

---

[2] The symplectic form is $-dq_1 \wedge dp_1 + dp_2 \wedge dq_2$. Because q_1 is circle-valued, the identification of q_1 with a coordinate on the fibre of T^*L_1 only makes sense if $q_1 \approx 0$.

can also check it directly by differentiating), but it is also exact: the obstruction to exactness[3] is the integral $\int \beta$ around the loop $p_1 = 0$, $q_2 \in [-\pi, \pi]$, that is,[4]

$$\int_{-\pi}^{\pi} \arctan\left(\frac{\sin q_2}{1 + \cos q_2}\right) dq_2 = 0.$$

Exactness means there is a function $\varphi(p_1, q_2)$ such that $\beta = d\varphi$. Pick $\epsilon > 0$ and let $\rho(p_1)$ be a cut-off function equal to 0 for $p_1 \leq -\ln 2 - \epsilon$ and equal to 1 for $-\ln 2 + \epsilon \leq p_1$. If we take $k_1 = (\ln 2 - \epsilon)^2$ and $k_2 = (\ln 2 + \epsilon)^2$ and define K_1, K_2 as we did earlier, then the Lagrangian cylinder given by the graph of $d(\rho\varphi)$ coincides with the cylinder L_1 outside K_2 and coincides with L in the compact region K_1.

We perform a similar modification near each of the three punctures. The result is a Lagrangian pair-of-pants which coincides with the three Lagrangian cylinders L_1, L_2, and L_3 outside the compact set K_2. This does not quite prove Theorem H.1, because we cannot take K_2 arbitrarily small in this construction. However, notice that the radial vector field in the p-plane is a Liouville vector field, by Lemma G.4. Our visible Lagrangian cylinders L_1, L_2, L_3 are preserved by the flow of this Liouville field, and if we flow L backwards along this Liouville field, we ensure that it agrees with L_1, L_2, L_3 on a larger and larger region (Figure H.3). This completes the proof of Theorem H.1. □

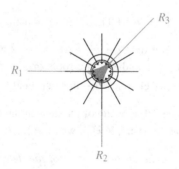

Figure H.3 We can find a Lagrangian whose projection to the p-plane is arbitrarily close to $R_1 \cup R_2 \cup R_3$ by flowing backwards along a Liouville field.

---

[3] On the cylindrical end $p_1 < -\ln 2$ whose de Rham cohomology has rank 1.

[4] This integral vanishes because the integrand is an odd function.

H.2 Immersed Lagrangians

Now suppose that in the statement of Theorem H.1, the primitive vectors v_1, v_2, v_3 still satisfy the balancing condition $v_1 + v_2 + v_3$, but that each pair fails to form a $\mathbb{Z}$-basis for $\mathbb{Z}^2$. If we write the matrix M whose rows are v_1 and v_2 then applying this matrix (on the right) gives an integer matrix sending $(1, 0)$ and $(0, 1)$ to v_1 and v_2 respectively. This map is not induced by a symplectic map $\mathbb{R}^2 \times T^2 \to \mathbb{R}^2 \times T^2$, but it is induced by a holomorphic covering map $h: \mathbb{C}^* \times \mathbb{C}^* \to \mathbb{C}^* \times \mathbb{C}^*$ of degree $\det(M)$, namely $h(\exp(p + iq)) = \exp(pM + iqM)$. The hyper-Kähler twist of $h(C)$ is another Lagrangian submanifold with three punctures asymptotic to Lagrangian cylinders living over the rays pointing in the v_1-, v_2-, and v_3-directions. As in Theorem H.1, we can modify this Lagrangian so that it actually coincides with these Lagrangian cylinders outside a compact set.

The main difference is that the resulting Lagrangian pair-of-pants is not embedded: it is only immersed.

Example H.3 Consider the case $v_1 = (2, -1)$, $v_2 = (-1, 2)$. We have

$$h(z_1, z_2) = (z_1^2/z_2, z_2^2/z_1).$$

A self-intersection of $h(C)$ corresponds to a pair of points $\xi, \xi' \in \mathbb{C}\backslash\{0, -1\}$ with $h(\xi, \xi + 1) = h(\xi', \xi' + 1)$. In this case, we can show there is precisely one self-intersection. Suppose that ξ and ξ' are distinct solutions to $h(z, z + 1) = (u, v)$ for some $u, v \in \mathbb{C}^*$. Then the quadratic equations

$$z^2 = u(z + 1), \quad (z + 1)^2 = vz$$

have ξ and ξ' as roots. But then $\xi + \xi' = -u = v - 2$ and $\xi\xi' = -u = 1$. In particular, $u = -1$ and $v = 3$ and ξ, ξ' are roots of $z^2 + z + 1$, so ξ and ξ' are $\frac{-1 \pm i\sqrt{3}}{2}$. Thus, there is precisely one self-intersection at $(-1, 3)$.

Write $|v \wedge w|$ for the absolute value of the determinant of the 2-by-2 matrix whose rows are v and w. In Example H.3, we have $|v_1 \wedge v_2| = 3$.

Theorem H.4 *Let Δ be the absolute value of the determinant of the matrix whose rows are v_1 and v_2. The number of self-intersections of the Lagrangian pair-of-pants is $\delta = \frac{\Delta-1}{2}$.*

Remark H.5 (Exercise H.13) If v_1, v_2, v_3 are primitive integer vectors with $v_1 + v_2 + v_3 = 0$ then $|v_k \wedge v_\ell|$ is an odd number and is independent of k, ℓ.

We will not prove Theorem H.4, and refer the interested reader to [80, Corollary 4.3]. We leave the following related lemma as an exercise.

Lemma H.6 (Exercise H.14) *Suppose we have several straight lines of rational slope in $\mathbb{R}^2$ incident on a point $b \in B$. Let $v_1, \ldots, k$ be primitive integer vectors pointing along these lines. Show that the visible Lagrangian cylinders above these lines have a total of $\delta(b)$ transverse intersections, where*

$$\delta(b) = \sum_{i<j} |v_i \wedge v_j|.$$

H.3 Lagrangians from Tropical Curves

We have already seen how to construct Lagrangian submanifolds living over straight lines of rational slope. Provided we are willing to allow pinwheel core and Schoen–Wolfson singularities, these straight lines are allowed to terminate on the toric boundary. Thanks to Lemma 6.15, we also have visible Lagrangian discs terminating on base-nodes of almost toric fibrations, providing they live over eigenlines. Now, courtesy of Theorem H.1 we have immersed Lagrangian pairs-of-pants living over trivalent vertices satisfying the balancing condition (H.1). Since this pair-of-pants coincides with the three visible Lagrangian cylinders over the edges of the graph (except in a small neighbourhood of the vertex), we can combine all of these constructions to get a Lagrangian submanifold (possibly singular and immersed) living over a trivalent graph whose edges have rational slope and whose vertices satisfy the balancing condition (H.1). These graphs are called *tropical curves* and the associated Lagrangians are called *tropical Lagrangians*. In summary, a tropical Lagrangian is made up of:

- an immersed pair-of-pants over every trivalent vertex b of the graph (with $\delta(b)$ self-intersections),
- a visible Lagrangian cylinder over every edge,
- a (p, q)-pinwheel core over every point where an edge terminates on an edge of the almost toric base diagram,
- a disc or Schoen–Wolfson cone over every point where an edge terminates at a vertex of the almost toric base diagram,
- a disc over every base-node at which an edge terminates, providing the edge points in the eigendirection for the affine monodromy of the base-node.

Rather than writing this carefully and formally, it is easier to explain in some examples.

Example H.7 Consider the tropical curve shown in Figure H.4(a). The three corners are all $\mathbb{Z}$-affine equivalent; for example, if we think of the trivalent

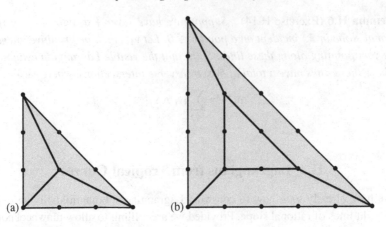

Figure H.4 (a) A tropical Lagrangian immersed sphere in $\mathbb{CP}^2$. (b) An embedded Lagrangian torus.

point as the origin then the matrix $\begin{pmatrix} -1 & 1 \\ -1 & 0 \end{pmatrix}$ sends the top vertex to the bottom left and preserves the tropical curve. The bottom left corner is the local model from Example 5.9, so the three corners give us visible Lagrangian discs with which to cap off the Lagrangian pair-of-pants coming from the 3-valent point. The outgoing vectors at the 3-valent point are:

$$v_1 = (-1, -1), \quad v_2 = (2, -1), \quad v_3 = (-1, 2)$$

so $|v_1 \wedge v_2| = \left| \det \begin{pmatrix} -1 & -1 \\ 2 & -1 \end{pmatrix} \right| = 3$, so the δ-invariant is $(3-1)/2 = 1$. This tropical Lagrangian is therefore an immersed sphere in $\mathbb{CP}^2$ with one transverse double point.

Example H.8   Consider the tropical curve shown in Figure H.4(b). The corners are all the same as in Example H.7. The trivalent vertices all have $\delta = 1$, so the result is an embedded Lagrangian. By inspection, it is topologically a torus.

Example H.9 (Exercise H.12)   Nemirovski and Shevchishin proved, independently and in very different ways, that there is no embedded Lagrangian Klein bottle in $\mathbb{CP}^2$. Why is Figure H.5 not a counterexample to their theorem?

Example H.10   The almost toric base diagrams in Figure H.6 represent blow-ups of the standard symplectic ball in three smaller balls, where the symplectic areas of the exceptional spheres after blowing up are a, b, and c. (You can get these diagrams by performing one toric blow-up at the origin and two

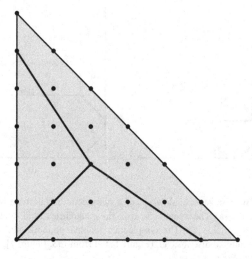

Figure H.5 Another tropical Lagrangian in $\mathbb{CP}^2$.

non-toric blow-ups and changing branch cuts.) In Figure H.6(a), the tropical Lagrangian associated with the tropical curve is a Lagrangian $\mathbb{RP}^2$ representing the $\mathbb{Z}/2$-homology class $E_1 + E_2 + E_3$, meeting the toric boundary along a $(2, 1)$-pinwheel core (Möbius strip). In Figure H.6(b), the tropical Lagrangian is a disc with boundary on the left-hand toric boundary. Shevchishin and Smirnov [98] showed that the $\mathbb{Z}/2$-homology class $E_1 + E_2 + E_3$ can be represented by a Lagrangian $\mathbb{RP}^2$ if and only if the *symplectic triangle inequalities* hold:

$$a < b + c, \quad b < c + a, \quad c < a + b.$$

Whenever these inequalities hold (e.g. Figure H.6(a)), the tropical curve modelled on this tripod furnishes us with a Lagrangian $\mathbb{RP}^2$, and whenever they fail, it furnishes us with a disc. For more tropical discussions along these lines, see [33].

H.4 Solutions to Inline Exercises

Exercise H.11 (Lemma H.2) *Consider the* hyper-Kähler twist

$$\mathbb{R}^2 \times T^2 \to \mathbb{R}^2 \times T^2, \quad (p_1, p_2, q_1, q_2) \mapsto (p_1, p_2, -q_2, q_1).$$

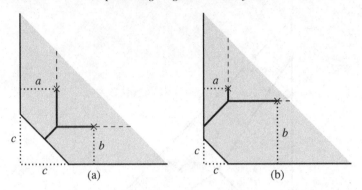

Figure H.6 Two almost toric blow-ups of a symplectic ball in three smaller balls (Example H.10). (a) The symplectic triangle inequalities hold, and we find a tropical Lagrangian $\mathbb{RP}^2$. (b) The symplectic triangle inequalities fail and we find a tropical Lagrangian disc instead. (Figure taken from [33, Figure 1], CC-BY 4.0 Jonathan David Evans.)

If $C \subseteq \mathbb{R}^2 \times T^2$ is a complex curve with respect to the complex coordinates $z_k = e^{p_k + iq_k}$ then show that the image of C under this hyper-Kähler twist is Lagrangian for the symplectic form $\sum dp_i \wedge dq_i$.

Solution Let $z = x + iy$ be a local complex coordinate on C and suppose $(p_1(z), p_2(z), q_1(z), q_2(z))$ is a local parametrisation of C, holomorphic with respect to the complex coordinates $z_k = e^{p_k + iq_k}$ Since (any branch of) the logarithm is holomorphic, this is equivalent to the Cauchy–Riemann equations

$$\frac{\partial p_k}{\partial x} = \frac{\partial q_k}{\partial y}, \quad \frac{\partial p_k}{\partial y} = -\frac{\partial q_k}{\partial x}, \text{ for } k = 1, 2.$$

After applying the twist, we get a submanifold which satisfies

$$\frac{\partial p_1}{\partial x} = \frac{\partial q_2}{\partial y}, \quad \frac{\partial p_1}{\partial y} = -\frac{\partial q_2}{\partial x}, \quad \frac{\partial p_2}{\partial x} = -\frac{\partial q_1}{\partial y}, \quad \frac{\partial p_2}{\partial y} = \frac{\partial q_1}{\partial x}.$$

This means that if ∂_x pushes forward along the twisted embedding to (a, b, c, d) then ∂_y pushes forward to $(-d, c, -b, a)$, and

$$\omega((a, b, c, d), (-d, c, -b, a)) = -ab + cd + ab - cd = 0,$$

so this twisted embedding is Lagrangian. $\qquad\qquad\qquad\qquad\qquad\qquad \square$

Exercise H.12 (Example H.9) *What is the tropical Lagrangian in Figure H.5?*

Solution Where the tropical curve meets the corner and edges, it gives us a Lagrangian disc and two Lagrangian Möbius strips. These are used to cap off the pair-of-pants coming from the trivalent vertex. Therefore, this tropical

Lagrangian is an immersed Klein bottle. The number of transverse double points is given by the δ-invariant of the trivalent vertex. The outgoing vectors at this point are:

$$v_1 = (-2, 3), \quad v_2 = (3, -2), \quad v_3 = (-1, -1),$$

so $|v_1 \wedge v_2| = \left| \det \begin{pmatrix} -2 & 3 \\ 3 & -2 \end{pmatrix} \right| = 5$ and $\delta = (5 - 1)/2 = 2$. Therefore, this Klein bottle has two double points.   □

Exercise H.13 (Remark H.5) *If v_1, v_2, v_3 are primitive integer vectors with $v_1 + v_2 + v_3 = 0$ then $|v_k \wedge v_\ell|$ is an odd number and is independent of k, ℓ.*

Solution If we switch the rows v_1 and v_2 then we just change the sign of the determinant. If we replace v_2 by v_3 then the balancing condition means $v_3 = -v_1 - v_2$, so

$$|v_1 \wedge v_3| = |v_1 \wedge (v_1 + v_2)| = |v_1 \wedge v_2|.$$

Similarly, the result is unchanged if we replace v_1 by v_3.

To see that it is odd, write $v_1 = (a, b)$ and $v_2 = (c, d)$. Suppose for contradiction that $|v_1 \wedge v_2| = |ad - bc|$ is even. We claim that a, b, c, d are then all odd. This will give a contradiction because it implies $v_3 = (-a - c, -b - d)$ is divisible by 2 and so not primitive.

To prove that $|ad - bc|$ even implies a, b, c, d are all odd, suppose that a is even (a similar argument works for b, c, d). Then $ad - bc = 0 \mod 2$ implies that either b or c is even. But if a is even then b is odd by primitivity of v_1, so c is even. This means d is odd. But then $v_3 = (-a - c, -b - d)$ has both components even, which contradicts primitivity.   □

Exercise H.14 (Lemma H.6) *Suppose we have several straight lines of rational slope in $\mathbb{R}^2$ incident on a point $b \in B$. Let $v_1, \dots, k$ be primitive integer vectors pointing along these lines. Show that the visible Lagrangian cylinders above these lines have a total of $\delta(b)$ transverse intersections, where*

$$\delta(b) = \sum_{i < j} |v_i \wedge v_j|.$$

Solution If $v_j = (m_j, n_j)$ for $j = 1, \dots, j$ then the visible Lagrangian cylinders intersect the fibre over b in the circles (in $\mathbb{R}^2/\mathbb{Z}^2$) with slopes $(-n_i, m_i)$. The number of intersections between two of these circles is $|m_2 n_1 - m_1 n_2|$. Summing this over pairs of circles gives $\delta(b)$.

For example, suppose the lines have slopes 1 and -1. Then the circles in the fibre have slopes -1 and 1 respectively. If we draw these as lines in the

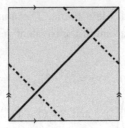

Figure H.7 Intersections between visible Lagrangian cylinders in the fibre over a point where their projections meet.

square with opposite sides identified (see Figure H.7), we see that these intersect twice. □

Appendix I

Markov Triples

I.1 The Markov Equation

The Diophantine equation

$$a^2 + b^2 + c^2 = 3abc$$

is called *the Markov equation*. It occurs in the theory of Diophantine approximation and continued fractions [12], the theory of quadratic forms [12], the study of exceptional collections [90], the hyperbolic geometry of a punctured torus [96], in governing the $\mathbb{Q}$-Gorenstein degenerations of $\mathbb{CP}^2$ [71, 51], and elsewhere; a wonderful exposition can be found in Aigner's book [2]. A triple of positive integers solving Markov's equation is called a *Markov triple*.

Lemma I.1 *If a, b, c is a Markov triple then so is $a, b, 3ab - c$.*

Proof Fix a and b and consider the quadratic function $f(x) := x^2 - 3abx + a^2 + b^2$. This quadratic has c as a root: $f(c) = 0$. A quadratic equation $x^2 + \beta x + \gamma = 0$ has two roots (counted with multiplicity) which sum to $-\beta$. In our case, $\beta = -3ab$, so this means that $3ab - c$ is another root. To see that both roots are positive, note that $f(0) = a^2 + b^2 > 0$ and $f'(0) = -3ab < 0$. This means that $f(x) > 0$ for all $x < 0$ (see Figure I.1) so there are no negative roots. □

Definition I.2 The operation of replacing the Markov triple (a, b, c) with $(a, b, 3ab - c)$ is called a *mutation* on c. The graph whose vertices are Markov triples and whose edges connect triples related by a mutation is called the *Markov graph*. In fact, this graph is a connected tree and we will usually refer to it as the *Markov tree*.

Theorem I.3 *The Markov graph is connected.*

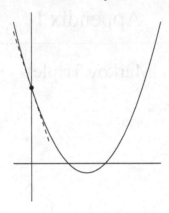

Figure I.1 The graph of f when $a = b = 1$. Positive y-intercept and negative gradient at $x = 0$ ensures both roots are positive.

Proof Suppose we are given a Markov triple (a, b, c) with $c > a, b$. Perform a mutation on the largest element c. This decreases the value of the largest element (Lemma I.4 below). If there is still a unique largest element, repeat this procedure. Since the values in the triple are decreasing, but always positive, this process terminates, and we find a Markov triple with a repeated largest element. The only triple with a repeated largest element is $(1, 1, 1)$ (Lemma I.5 below). This shows that the graph is connected. □

Lemma I.4 *Let a, b, c be a Markov triple with $a \le b \le c$. Then b lies in the closed interval between c and $3ab - c$. If $a < b$ then b lies in the interior of this interval. In particular, if $b < c$ then $3ab - c \le b$.*

Proof If $f(x) = x^2 - 3abx + a^2 + b^2$ then $f(b) = b^2 - 3ab^2 + a^2 + b^2 = (2 - 3a)b^2 + a^2 \le a^2 - b^2 \le 0$. This means that b lies in the region between the two roots of f, and strictly between if $a^2 - b^2 < 0$. These roots are c and $3ab - c$. □

Lemma I.5 *The only Markov triple with no unique largest element is $(1, 1, 1)$.*

Proof If (a, b, c) is a Markov triple with $a \le b = c$ then substituting $b = c$ in the Markov equation gives $a^2 + 2b^2 = 3ab^2$, or $a^2 = (3a - 2)b^2$. Since $a \ge 1$, $3a - 2 \ge 1$ and hence $a^2 \ge b^2$. Since $a \le b$, we get $a = b = c$. The Markov equation now reduces to $a^2 = (3a - 2)a^2$, so $3a - 2 = 1$ and $a = b = c = 1$. □

Remark I.6 We briefly recall the definition of a tree. A *path* in a graph is a sequence of oriented edges such that the endpoint of edge i is the start-point of edge $i + 1$ for all i. A path is *non-simple* if the sequence of edges contains

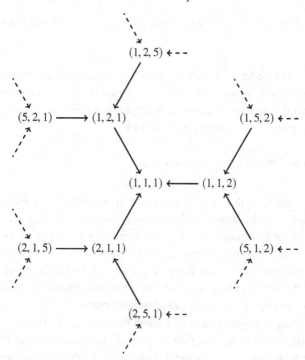

Figure I.2 A small part of the Markov tree (dashed lines indicate it continues). The arrows are explained in Lemma I.7.

a subsequence of the form $e\bar{e}$ where e is an edge and $\bar{e}$ is the same edge with the opposite orientation. Otherwise, a path is called *simple*. A graph is a *tree* if any pair of vertices can be connected by a unique simple path.

Lemma I.7 *The following prescription defines a global choice of orientations on the edges of the Markov graph. If an edge connects two triples (a, b, c) and $(a, b, 3ab - c)$ then orient it so that it points towards the triple whose maximal element is smaller. For example $(1, 1, 2) \rightarrow (1, 1, 1)$.*

Proof To show this is well-defined, we need to show that the maximal elements in (a, b, c) and $(a, b, 3ab - c)$ are different (otherwise there is no way to decide the orientation of the edge). Without loss of generality, suppose $a \le b$ and $c < 3ab - c$. Lemma I.4 tells us that $c \le b \le 3ab - c$. Therefore, $\max(a, b, c) = b$ and $\max(a, b, 3ab - c) = 3ab - c$. The only problem is if $b = 3ab - c$. This is only possible if $b = a$ by Lemma I.4, but in this case the triple (a, b, c) has a repeated maximum $a = b$, so $(a, b, c) = (1, 1, 1)$ and all the neighbouring triples are permutations of $(1, 1, 2)$, so it is clear how to orient the graph around this vertex.   □

Remark I.8   The vertex $(1, 1, 1)$ has three incoming edges. The proof of Lemma I.7 shows that every other vertex has two incoming edges (corresponding to mutations on the smaller elements of the triple) and one outgoing edge (corresponding to mutation on the largest element).

Theorem I.9   *The Markov graph is a tree.*

Proof   We have seen that the Markov graph is connected. It remains to show that two triples can be connected by a *unique* simple path. Orient the Markov graph according to Lemma I.7. We call a path *downwards* if it follows the orientation. The proof of Theorem I.3 can be interpreted as saying that there is a unique downwards path from every vertex to $(1, 1, 1)$. Define the *height* of a triple to be the length of this unique downwards path. Similarly, we can define the height of an edge to be the height of its start-point.

Fix two Markov triples t_1 and t_2. Follow the paths down from t_1 and from t_2. Eventually these paths meet up at a vertex (this could be at $(1, 1, 1)$ or somewhere higher up). Let m be the first vertex where these paths meet; we get a simple path P by going down from t_1 to m and then up from m to t_2.

Suppose there is another simple path, Q. If $P \neq Q$ then there is an edge in Q which does not appear in P. Amongst the edges of Q which do not appear in P, let e_i be one which maximises height. There are three possible cases, each leading to a contradiction:

- Suppose $e_i = e_1$ is the first edge in Q. Since it is not in P, it cannot be the downward arrow from t_1, so it must go up. Since it is a highest edge, the next edge e_2 must return along $\bar{e}_1$ as there is only one way down. This contradicts the fact that Q is simple.

- A similar argument applies when e_i is the final edge in Q.

- If e_i is neither the first nor last edge in Q then there are edges e_{i-1} and e_{i+1} adjacent in the path. At least one of these must start at the highest point of e_i. Since it cannot go higher, it must be $\bar{e}_i$, which contradicts the assumption that Q is simple.

Therefore, there is a unique simple path between two vertices and the Markov graph is a tree.   □

I.2 Triangles

A *Vianna triangle* is an almost toric diagram whose edges are v_1, v_2, v_3 with affine lengths ℓ_1, ℓ_2, ℓ_3 and whose vertices P_1, P_2, P_3 are modelled on the T-singularities $\frac{1}{d_k p_k^2}(1, d_k p_k q_k - 1)$ for $k = 1, 2, 3$. The relative orientations and positions of the edges and vertices are shown in the following diagram. We call the numbers d_k, p_k, q_k, ℓ_k, $k = 1, 2, 3$, the *Vianna data* of the triangle.

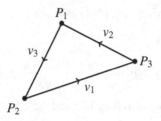

Our indices take values in the cyclic group $\mathbb{Z}/3$.

Lemma I.10 *If $\hat{v}_k$ denotes the primitive integer vector along v_k and ℓ_k denotes the affine length of v_k then we have the following relation:*[1]

$$\hat{v}_k \wedge \hat{v}_{k+1} = d_{k+2} p_{k+2}^2.$$

Proof Since the vertex P_{k+2} is modelled on the moment polygon of the $\frac{1}{d_{k+2} p_{k+2}^2}(1, d_{k+2} p_{k+2} q_{k+2} - 1)$ singularity, there is an integral affine transformation making $\hat{v}_k = (0, -1)$ and $\hat{v}_{k+1} = (d_{k+2} p_{k+2}^2, d_{k+2} p_{k+2} q_{k+2} - 1)$. Thus $\hat{v}_k \wedge \hat{v}_{k+1} = (0, -1) \wedge (d_{k+2} p_{k+2}^2, d_{k+2} p_{k+2} q_{k+2} - 1) = d_{k+2} p_{k+2}^2$. □

Corollary I.11 *We have*

$$\ell_1 \ell_2 d_3 p_3^2 = \ell_2 \ell_3 d_1 p_1^2 = \ell_3 \ell_1 d_2 p_2^2.$$

Proof Since $v_1 + v_2 + v_3 = 0$, we get

$$0 = (v_1 + v_2 + v_3) \wedge v_k = v_{k-1} \wedge v_k + 0 + v_{k+1} \wedge v_k = v_{k-1} \wedge v_k - v_k \wedge v_{k+1}.$$

The quantity $v_k \wedge v_{k+1}$ is therefore independent of k. Since $v_k = \ell_k \hat{v}_k$ and $v_{k+1} = \ell_{k+1} \hat{v}_{k+1}$, the previous lemma implies this is $\ell_k \ell_{k+1} d_{k+2} p_{k+2}^2$. □

Definition I.12 We write K for the common value $\ell_k \ell_{k+1} d_{k+2} p_{k+2}^2$ and L for the total affine length $\ell_1 + \ell_2 + \ell_3$.

Corollary I.13 *We have* $\ell_k = \dfrac{p_k}{p_{k+1} p_{k+2}} \sqrt{\dfrac{K d_k}{d_{k+1} d_{k+2}}}$.

---

[1] Recall that $(a, b) \wedge (c, d) := ad - bc$.

Proof   For concreteness, take $k = 3$. By Corollary I.11, we have

$$\ell_1 = \frac{K}{\ell_3 d_2 p_2^2}, \quad \ell_2 = \frac{K}{\ell_3 d_1 p_1^2}, \quad \ell_1 \ell_2 d_3 p_3^2 = K,$$

so

$$\frac{K^2 d_3 p_3^2}{\ell_3^2 d_1 d_2 p_1^2 p_2^2} = K,$$

which gives $\ell_3^2 = (K d_3 / d_1 d_2)(p_3^2 / p_1^2 p_2^2)$ as required.   □

Corollary I.14   *We have*

$$d_1 p_1^2 + d_2 p_2^2 + d_3 p_3^2 = \frac{L \sqrt{d_1 d_2 d_2}}{\sqrt{K}} p_1 p_2 p_3.$$

Proof   This follows from Corollary I.13 and the fact that $\ell_1 + \ell_2 + \ell_3 = L$.   □

We will now prove a sequence of lemmas to show that the constants K and L are unchanged by mutation.

Lemma I.15   *The eigenline at vertex P_{k+2} points in the direction $\frac{\hat{v}_{k+1} - \hat{v}_k}{d_{k+2} p_{k+2}}$.*

Proof   Again, making an integral affine transformation we can assume $\hat{v}_k = (0, -1)$ and $\hat{v}_{k+1} = (d_{k+2} p_{k+2}^2, d_{k+2} p_{k+2} q_{k+2} - 1)$. In these coordinates, the eigenline points in the (p_{k+2}, q_{k+2})-direction, which is $(\hat{v}_{k+1} - \hat{v}_k)/d_{k+2} p_{k+2}$.   □

Lemma I.16   *If we perform a mutation on the vertex P_3 then we obtain a new Vianna triangle with data (omitting the q_k's):*

$$d_1' = d_1, \qquad\qquad d_2' = d_2, \qquad\qquad d_3' = d_3,$$

$$p_1' = p_1, \qquad\qquad p_2' = p_2, \qquad\qquad p_3' = (d p_1^2 + d_2 p_2^2)/d_3 p_3,$$

$$\ell_1' = \frac{\ell_3 d_2 p_2^2}{d_1 p_1^2 + d_2 p_2^2}, \quad \ell_2' = \frac{\ell_3 d_1 p_1^2}{d_1 p_1^2 + d_2 p_2^2}, \quad \ell_3' = \ell_1 + \ell_2.$$

Proof   When we mutate, a new vertex is introduced at the point P_3' where the edge v_3 intersects the eigenline emanating out of P_3 in the $\frac{\hat{v}_2 - \hat{v}_1}{d_3 p_3}$ direction. We find the new p_3' by taking $\frac{\hat{v}_2 - \hat{v}_1}{d_3 p_3} \wedge \hat{v}_3$, which gives

$$p_3' = \frac{1}{d_3 p_3}(\hat{v}_2 \wedge \hat{v}_3 - \hat{v}_1 \wedge \hat{v}_3) = \frac{d_1 p_1^2 + d_2 p_2^2}{d_3 p_3}.$$

This intersection point P_3' lives on the line joining P_1 and P_2. If we choose coordinates where P_3 is the origin, P_1 and P_2 are the vectors v_2 and $-v_1$ respectively. Let $s \in [0, 1]$ be the number such that $P_3 = -s v_1 + (1 - s) v_2$. This lives on the eigenline, so can be written as $t(\hat{v}_2 - \hat{v}_1)$ for some t. Since v_1 and

v_2 are linearly independent, we deduce that $-sv_1 = -t\hat{v}_1$ and $(1-s)v_2 = t\hat{v}_2$. Eliminating t gives $s/(1-s) = \ell_2/\ell_1$. We have $\ell_2/\ell_1 = d_2p_2^2/d_1p_1^2$ by Corollary I.13, so we get $s = \frac{d_2p_2^2}{d_1p_1^2+d_2p_2^2}$. Thus $P_3' = \frac{d_1p_1^2v_2-d_2p_2^2v_1}{d_1p_1^2+d_2p_2^2}$. The vector from $P_1 = v_2$ to P_3' is therefore $-\frac{d_2p_2^2}{d_1p_1^2+d_2p_2^2}(v_1+v_2) = \frac{d_2p_2^2}{d_1p_1^2+d_2p_2^2}v_3$. Thus, the edge v_3 is subdivided so that its affine length ℓ_3 is split in the ratio $d_1p_1^2 : d_2p_2^2$. $\qquad\square$

Corollary I.17 *The common value* $\ell_1\ell_2d_3p_3^2 = \ell_2\ell_3d_1p_1^2 = \ell_3\ell_1d_2p_2^2$ *is unchanged by mutation at P_3 (or, by symmetry, any other vertex). Similarly, the sum $\ell_1 + \ell_2 + \ell_3$ is unchanged by mutation.*

Proof We compute the value $\ell_1'\ell_2'd_3'(p_3')^2$ after mutation:

$$\ell_1'\ell_2'd_3'(p_3')^2 = \frac{\ell_3d_2p_2^2 \cdot \ell_3d_1p_1^2}{(d_1p_1^2+d_2p_2^2)^2}\, d_3 \left(\frac{d_1p_1^2+d_2p_2^2}{d_2p_3}\right)^2 = \frac{\ell_3^2d_1p_1^2d_2p_2^2}{d_3p_3^2}.$$

By Corollary I.11, $\ell_3d_1p_1^2 = \ell_1d_3p_3^2$ and $\ell_3d_2p_2^2 = \ell_2d_3p_3^2$, so this reduces to $\ell_1\ell_2d_3p_3^2$, which is the same as before mutation.

The sum of the affine lengths after mutation is

$$(\ell_1 + \ell_2) + \frac{\ell_3d_2p_2^2}{d_1p_1^2+d_2p_2^2} + \frac{\ell_3d_1p_1^2}{d_1p_1^2+d_2p_2^2} = \ell_1 + \ell_2 + \ell_2. \qquad\square$$

Example I.18 (Proof of Theorem 8.21) The triangle $D(1,1,1)$ in Figure I.3 is a Vianna triangle with data $d_1 = d_2 = d_3 = 1$, $p_1 = p_2 = p_3 = 1$, and $\ell_1 = \ell_2 = \ell_3 = 3$. This has $K = 9$ and $L = 9$, so Corollaries I.14 and I.17 tell us that if D is obtained from $D(1,1,1)$ by iterated mutation and has Vianna data d_k, p_k, ℓ_k then

$$p_1^2 + p_2^2 + p_3^2 = 3p_1p_2p_3$$

and $\ell_k = 3p_k/(p_{k+1}p_{k+2})$. Moreover, if we write $D(p_1, p_2, p_3)$ for the triangle associated with the Markov triple p_1, p_2, p_3, mutation at vertex 3 gives the Markov triple $p_1, p_2, p_3' = 3p_1p_2 - p_3$. One can check this from the formulas, or simply observe that this mutation leaves p_1 and p_2 unchanged, and there are only two Markov triples containing both p_1 and p_2.

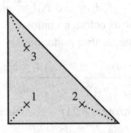

Figure I.3 The almost toric diagram $D(1, 1, 1)$. The edges all have affine length 3.

Appendix J

Open Problems

In the preface, I mentioned that this book is intended to give you the tools to explore further in symplectic geometry. In this final appendix, you will find a handful of open[1] problems which you might like to explore. These are some of my pet problems: not particularly venerable or widely known, but they have intrigued and frustrated me over the years. The first few are hopefully purely 'combinatorial' or geometric in that they should not require any Floer theory to solve. Some will probably require Floer theory, and I have not hesitated to use terminology from beyond this book in discussing these problems, because you will need to learn it to solve them.

J.1 Mutation of Quadrilaterals

Problem J.1 Which quadrilaterals arise as mutations of a square? Or a rectangle? For a rectangle, how does the ratio of the side-lengths affect the answer?

It is easy to generate such quadrilaterals by iterated mutation, but I know of no succinct description of this class of quadrilaterals analogous to the Markov-triple description of mutants of the $\mathbb{CP}^2$ triangle.

J.2 Fillings of Lens Spaces

Equip the lens space $L(n, a)$ with its standard contact structure. It is known that $L(n, a)$ has a filling with second Betti number zero if and only if $n = p^2$ and $a = pq - 1$ for some coprime integers p, q; moreover, such a filling is unique up to symplectomorphism. It is also known ([52, Theorem 4.3], [109, Theorem 2.8])

---

[1] At the time of writing.

that $L(n, a)$ admits at most two fillings (up to deformation/symplectomorphism) with $b_2 = 1$. For example, in Remark 9.8 we saw that $L(36, 13)$ admits two non-diffeomorphic fillings.

Problem J.2   Characterise which (standard contact) lens spaces admit precisely two symplectic fillings with $b_2 = 1$.

This is a purely combinatorial problem, thanks to Lisca's classification. Again, it is easy to generate examples (even infinite families of examples) and there are many tantalising patterns, but I do not know a complete answer. These examples arise very naturally in the context of algebraic geometry; the papers [108] and [109] have many examples of surfaces of general type containing a lens space hypersurface which are related (topologically) by the surgery which interchanges the two fillings. Urzúa and Vilches [109] call these *wormholes* in the moduli space of surfaces.

J.3 Minimal Genus Problem

Let (X, ω_λ) be the symplectic manifold $S^2 \times S^2$ with a product symplectic form giving the factors areas 1 and λ. Let $\beta \in H_2(S^2 \times S^2; \mathbb{Z}/2)$ be the homology class of $S^2 \times \{pt\}$.

Problem J.3   As a function of λ, what is the minimal genus of a (nonorientable) embedded Lagrangian submanifold in X inhabiting the $\mathbb{Z}/2$-homology class β? Is this genus uniformly bounded in λ?

Here, the genus of the nonorientable surface $\#_m \mathbb{RP}^2$ is defined to be m. One can find [33] explicit tropical Lagrangians in this homology class whose genus goes to infinity with m, and it seems difficult (to me) to do significantly better than this with tropical Lagrangians.

Proving a lower bound on the minimal genus will be difficult and most likely require techniques of Floer theory, but you can give an upper bound just by constructing examples.

Obviously this problem admits many generalisations (just pick a different manifold X) but this seems to be the simplest version which is open.

J.4 Nodal Slides in Higher Dimensions

In this book, we focused on symplectic 4-manifolds fibring over two-dimensional integral affine bases. Almost toric fibrations make sense in higher dimensions too, and the discriminant locus has codimension 2. For example, if the base is three-dimensional then the discriminant locus will be some kind of knotted graph.

A nodal slide is a kind of isotopy of the discriminant locus. If the base is two-dimensional, you can usually slide base-nodes out of the way of another sliding node[2] which gives great flexibility in constructing and modifying Lagrangian torus fibrations. But in dimension 3, this is no longer true: in a generic one-parameter family of Lagrangian torus fibrations, you expect a discrete set of fibrations where the discriminant locus fails to be embedded. It is not clear how to continue the nodal slide beyond this point.

Problem J.4 Is there a theory of nodal slides in dimension 3? If the discriminant locus hits itself at some point along the slide, can we modify the fibration to allow the discriminant locus to pass through itself?

See the work of Groman and Varolgunes [43] for a different and enlightening perspective on nodal slides.

J.5 Lagrangian Rational Homology Spheres

Lagrangian submanifolds L with $H^1(L; \mathbb{R}) = 0$ are 'rigid' in the sense that any family of Lagrangian submanifolds L_t with $L_0 = L$ come from a Hamiltonian isotopy in the sense that $L_t = \phi_t^{H_t}(L)$ for some time-dependent Hamiltonian H_t. If L is three-dimensional then $H^1(L; \mathbb{R}) = 0$ if and only if L is a *rational homology sphere*, that is, $H^*(L; \mathbb{Q}) \cong H^*(S^3; \mathbb{Q})$.

In $\mathbb{CP}^3$, we have two well-known Lagrangian rational homology spheres: $\mathbb{RP}^3$ and the *Chiang Lagrangian* (see [18, 34, 103]).

Problem J.5 Are there any other Lagrangian rational homology 3-spheres in $\mathbb{CP}^3$?

Konstantinov [60] used Floer theory to give strong restrictions on which homology spheres could occur. Namely, in [60, Corollary 3.2.12], he showed that the only possibility is that L is a quotient of S^3 by a finite subgroup of $SO(4)$ which is either cyclic of order $4k$ for some $k \geq 1$ or else isomorphic to a product $D_{2^k(2n+1)} \times C_m$ where $k \geq 2$, $n \geq 1$, $\gcd(2^k(2n+1), m) = 1$ and D_ℓ is the dihedral group of order ℓ. The case $D_{12} \times C_1$ is the Chiang Lagrangian.

Ruling the other cases out would probably require Floer theoretic machinery, but maybe one can construct examples using tropical techniques: Mikhalkin's

---

[2] Unless their eigenlines are collinear, in which case they can harmlessly slide over one another anyway because the vanishing cycles can be made disjoint.

theory works for tropical curves in any dimension. In particular, Mikhalkin has a construction of a Chiang-like Lagrangian [80, Example 6.20] and he raises Problem J.5 as [80, Question 6.22].

J.6 Disjoint Pinwheels

Evans and Smith [36] showed that one can find at most three pairwise-disjointly embedded Lagrangian pinwheels in $\mathbb{CP}^2$. The argument is somewhat convoluted and mysterious; in particular, we *first* showed that any triple of disjoint pinwheels satisfy Markov's equation, and then deduced from this that there can be at most three. Since we do not know what the analogue of the Markov equation should be for other Del Pezzo surfaces like $S^2 \times S^2$, it is not clear how to generalise the result.

Our result was inspired by a theorem of Hacking and Prokhorov [51] which bounded the number of cyclic quotient singularities on a singular degeneration of $\mathbb{CP}^2$. Prokhorov has a more general bound for the number of cyclic quotient singularities in a $\mathbb{Q}$-Gorenstein degeneration of a Del Pezzo surface in terms of the Picard rank ρ of the singular fibre (there should be at most $\rho + 2$). This in turn is bounded above by the rank of the quantum cohomology of the smooth fibre, which motivates the following question:

Problem J.6 Let X be a monotone symplectic 4-manifold with semisimple quantum cohomology[3] and let r be the rank of its quantum cohomology. Is it true that there are at most r pairwise-disjointly embedded Lagrangian pinwheels in X?

Upper bounds like this hold for pairwise-disjointly embedded Lagrangian submanifolds whose Floer cohomology is non-vanishing over some field when the quantum cohomology is semisimple over that field, see [32, Theorem 1.25], [93, Theorem 1.3]. However, Lagrangian pinwheels do not have well-defined Floer cohomology, and even if they did it would be likely that one must work over fields of different characteristics to get well-defined Floer cohomology (even to get a fundamental class). Perhaps one can make some headway using recent developments in relative symplectic cohomology due to McLean [79], Varolgunes [113], and Venkatesh [114]?

---

[3] See [76] for an introduction to quantum cohomology, and the references in the following paragraph for what it might have to do with disjoint Lagrangian embeddings. Technically, we should specify which ground field we are working over for this assumption to make sense.

J.7 Pinwheel Content and Deformations of ω

Given a symplectic 4-manifold (X, ω), define its *pinwheel content* to be the set of pairs (p, q) such that some neighbourhood of the Lagrangian pinwheel in $B_{1,p,q}$ embeds symplectically in X. Combining Problems J.1, J.3, and J.6, we can ask how the pinwheel content depends on the cohomology class of ω. The simplest example is $X = S^2 \times S^2$ with the symplectic form ω_λ giving the factors areas 1 and λ. For $\lambda < 2$, it is possible to construct a symplectically embedded $B_{1,3,1} \subseteq X$: this becomes visible after mutating the standard moment rectangle (see Figure J.1(a)). For $\lambda \geq 2$, this construction fails (Figure J.1(b)).

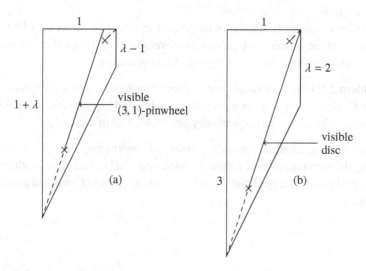

Figure J.1 (a) A $(3, 1)$-pinwheel in $(S^2 \times S^2, \omega_\lambda)$ for $\lambda < 2$. (b) What becomes of this when $\lambda \geq 2$.

Problem J.7   Is there a symplectically embedded $B_{1,3,1} \subseteq S^2 \times S^2$ when $\lambda \geq 2$?

A much harder open question along these lines (about symplectic embeddings of $B_{1,p,q}$s in surfaces of general type) was raised in [38, Remark 1.4].

J.8 Big Balls

The final problem I want to state is motivated by a result of Ein, Küchle, and Lazarsfeld [28] which gives a lower bound on the Seshadri constants of projective varieties. Rather than explaining this theorem, here is how Lazarsfeld [62, Remark 5.2.7] reformulates their result in terms of symplectic topology:

Theorem J.8 *Let $X \subseteq \mathbb{CP}^N$ be a smooth projective variety of complex dimension n and let ω be the symplectic form on X pulled back from the Fubini–Study form on $\mathbb{CP}^N$. Then X contains a symplectically embedded ball of radius $1/\sqrt{n\pi}$.*

All symplectic manifolds contain symplectic balls of some (possibly very small) radius, but this result is saying you can always find a 'big ball' (of size at least $1/\sqrt{n\pi}$) if X is the symplectic manifold underlying a complex n-dimensional projective variety.

The proof makes heavy use of techniques in algebraic geometry which are not available in symplectic topology. However, one might hope to find a purely symplectic construction of this big ball. More generally:

Problem J.9 (Big Ball conjecture)   Prove that there exists a universal constant R_n depending only on n such that any integral[4] symplectic manifold of dimension $2n$ admits a symplectically embedded ball of radius R_n.

Almost toric fibrations provide a means of constructing balls, for example taking the preimage of a suitable neighbourhood of a Delzant vertex. Schlenk's book [91] provides a wealth of material on how to construct balls and packings by balls.

---

[4] 'Integral' here means $[\omega]/2\pi$ is an integral cohomology class. You might want to think about why this condition is necessary.

References

[1] B. Acu, O. Capovilla-Searle, A. Gadbled, A. Marinkovic, E. Murphy, L. Starkston, and A. Wu. An introduction to Weinstein handlebodies for complements of smoothed toric divisors. In B. Acu, C. Cannizzo, D. McDuff, Z. Myer, Y. Pan, and L. Traynor editors, *Research directions in symplectic and contact geometry and topology*, volume 27 of Association for Women in Mathematics Series, pages 217–43. Springer, Cham, 2021.

[2] M. Aigner. *Markov's theorem and 100 years of the uniqueness conjecture: A mathematical journey from irrational numbers to perfect matchings*. Springer, Cham, 2013.

[3] V. I. Arnold. *Mathematical methods of classical mechanics*, volume 60 of Graduate Texts in Mathematics. Springer, New York, 1989. Translated from the 1974 Russian original by K. Vogtmann and A. Weinstein, 2nd ed.

[4] M. F. Atiyah. Convexity and commuting Hamiltonians. *Bull. London Math. Soc.*, 14(1):1–15, 1982.

[5] M. Audin. *Torus actions on symplectic manifolds*, volume 93 of Progress in Mathematics. Birkhäuser, Basel, revised ed., 2004.

[6] M. Audin. Lagrangian skeletons, periodic geodesic flows and symplectic cuttings. *Manuscripta Math.*, 124(4):533–50, 2007.

[7] D. Auroux. Mirror symmetry and T-duality in the complement of an anticanonical divisor. *J. Gökova Geom. Topol. GGT*, 1:51–91, 2007.

[8] M. Bertozzi, T. Holm, E. Maw, D. McDuff, G. T. Mwakyoma, A. R. Pires, and M. Weiler. Infinite staircases for Hirzebruch surfaces. In B. Acu, C. Cannizzo, D. McDuff, Z. Myer, Y. Pan, and L. Traynor editors, *Research directions in symplectic and contact geometry and topology*, volume 27 of Association for Women in Mathematics Series, pages 47–157. Springer, Cham, 2021.

[9] M. Bhupal and K. Ono. Symplectic fillings of links of quotient surface singularities. *Nagoya Math. J.*, 207:1–45, 2012.

[10] R. Bott and L. W. Tu. *Differential forms in algebraic topology*, volume 82 of Graduate Texts in Mathematics. Springer, Berlin, 1982.

[11] R. Casals and R. Vianna. Full ellipsoid embeddings and toric mutations. *Selecta Math. (N.S.)*, 28(3): Paper No. 61, 2022.

[12] J. W. S. Cassels. *An introduction to Diophantine approximation*, volume 45 of Cambridge Tracts in Mathematics and Mathematical Physics. Cambridge University Press, New York, 1957.

[13] R. Castaño-Bernard and D. Matessi. Lagrangian 3-torus fibrations. *J. Differential Geom.*, 81(3):483–573, 2009.

[14] R. Castaño-Bernard and D. Matessi. Semi-global invariants of piecewise smooth Lagrangian fibrations. *Q. J. Math.*, 61(3):291–318, 2010.

[15] M. Chaperon. Normalisation of the smooth focus–focus: A simple proof. *Acta Math. Vietnam.*, 38(1):3–9, 2013.

[16] Yu. V. Chekanov. Lagrangian tori in a symplectic vector space and global symplectomorphisms. *Math. Z.*, 223(4):547–59, 1996.

[17] M.-W. M. Cheung and R. Vianna. Algebraic and symplectic viewpoint on compactifications of two-dimensional cluster varieties of finite type. In D. R. Wood, J. de Gier, C. E. Praeger, and T. Tao, editors, *2019–20 MATRIX Annals*, volume 4 of MATRIX Book Series, pages 567–602. Springer, Cham, 2021.

[18] R. Chiang. New Lagrangian submanifolds of $\mathbb{CP}^n$. *Int. Math. Res. Not.*, (45):2437–41, 2004.

[19] J. A. Christophersen. On the components and discriminant of the versal base space of cyclic quotient singularities. In D. Mond and J. Montaldi, editors, *Singularity theory and its applications, Part I (Coventry, 1988/1989)*, volume 1462 of Lecture Notes in Mathematics, pages 81–92. Springer, Berlin, 1991.

[20] M. T. Cicero. *Tusculanae Disputationes V*. Loeb Classical Library. Harvard University Press, Cambridge, MA, 1927.

[21] J. H. Conway. *The sensual (quadratic) form*, volume 26 of Carus Mathematical Monographs. Mathematical Association of America, Washington, DC, 1997. With the assistance of Francis Y. C. Fung.

[22] D. Cristofaro-Gardiner, T. Holm, A. Mandini, and A. R. Pires. On infinite staircases in toric symplectic four-manifolds. *arXiv:2004.07829*, 2020.

[23] V. I. Danilov. The geometry of toric varieties. Russian Mathematical Surveys, 33(2):97–154, 1978.

[24] T. Delzant. Hamiltoniens périodiques et images convexes de l'application moment. *Bull. Soc. Math. France*, 116(3):315–39, 1988.

[25] J.-P. Dufour and P. Molino. Compactification d'actions de $\mathbf{R}^n$ et variables action-angle avec singularités. In P. Dazord and A. Weinstein, editors, *Séminaire Sud-Rhodanien 1ère partie*, volume 1988, issue 1B of Publications du Département de Mathématiques de Lyon, pages 161–83. 1988.

[26] J. J. Duistermaat. On global action-angle coordinates. *Comm. Pure Appl. Math.*, 33(6):687–706, 1980.

[27] J. Ebert. Relative de rham cohomologies. MathOverflow. https://mathoverflow.net/q/111063 (version: 2012-10-30).

[28] L. Ein, O. Küchle, and R. Lazarsfeld. Local positivity of ample line bundles. *J. Differential Geom.*, 42(2):193–219, 1995.

[29] Y. Eliashberg. Contact 3-manifolds twenty years since J. Martinet's work. *Ann. Inst. Fourier (Grenoble)*, 42(1–2):165–92, 1992.

[30] L. H. Eliasson. Normal forms for Hamiltonian systems with Poisson commuting integrals – elliptic case. *Comment. Math. Helv.*, 65(1):4–35, 1990.

[31] P. Engel. Looijenga's conjecture via integral-affine geometry. *J. Differential Geom.*, 109(3):467–95, 2018.

[32] M. Entov and L. Polterovich. Rigid subsets of symplectic manifolds. *Compos. Math.*, 145(3):773–826, 2009.

[33] J. D. Evans. A Lagrangian Klein bottle you can't squeeze. *J. Fixed Point Theory Appl.*, 24(2): Paper No. 47, 2022.

[34] J. D. Evans and Y. Lekili. Floer cohomology of the Chiang Lagrangian. *Selecta Math. (N.S.)*, 21(4):1361–404, 2015.

[35] J. D. Evans and M. Mauri. Constructing local models for Lagrangian torus fibrations. *Ann. H. Lebesgue*, 4:537–70, 2021.

[36] J. D. Evans and I. Smith. Markov numbers and Lagrangian cell complexes in the complex projective plane. *Geom. Topol.*, 22(2):1143–80, 2018.

[37] J. D. Evans and I. Smith. Bounds on Wahl singularities from symplectic topology. *Algebr. Geom.*, 7(1):59–85, 2020.

[38] J. D. Evans and G. Urzúa. Antiflips, mutations, and unbounded symplectic embeddings of rational homology balls. *Ann. Inst. Fourier (Grenoble)*, 71(5):1807–43, 2021.

[39] R. Fintushel and R. J. Stern. Rational blowdowns of smooth 4-manifolds. *J. Differential Geom.*, 46(2):181–235, 1997.

[40] W. Fulton. *Introduction to toric varieties*, volume 131 of Annals of Mathematics Studies. Princeton University Press, Princeton, NJ, 1993.

[41] R. E. Gompf. A new construction of symplectic manifolds. *Ann. of Math. (2)*, 142(3):527–95, 1995.

[42] A. G. Greenhill. *The applications of elliptic functions*. Dover, Mineola, NY, 1959.

[43] Y. Groman and U. Varolgunes. Locality of relative symplectic cohomology for complete embeddings. *arXiv:2110.08891*, 2021.

[44] M. Gromov. Pseudo holomorphic curves in symplectic manifolds. *Invent. Math.*, 82(2):307–47, 1985.

[45] M. Gross. Examples of special Lagrangian fibrations. In K. Fukaya, Y.-G. Oh, K. Ono, and G. Tian, editors, *Symplectic geometry and mirror symmetry (Seoul, 2000)*, pages 81–109. World Scientific, River Edge, NJ, 2001.

[46] M. Gross. Special Lagrangian fibrations. I. Topology. In C. Vafa and S.-T. Yau, editors, *Winter School on Mirror Symmetry, Vector Bundles and Lagrangian Submanifolds (Cambridge, MA, 1999)*, volume 23 of AMS/IP Studies in Advanced Mathematics, pages 65–93. American Mathematical Society, Providence, RI, 2001.

[47] M. Gross. Special Lagrangian fibrations. II. Geometry. A survey of techniques in the study of special Lagrangian fibrations. In C. Vafa and S.-T. Yau, editors, *Winter School on Mirror Symmetry, Vector Bundles and Lagrangian Submanifolds (Cambridge, MA, 1999)*, volume 23 of AMS/IP Studies in Advanced Mathematics, pages 95–150. American Mathematical Society, Providence, RI, 2001.

[48] M. Gross, P. Hacking, and S. Keel. Mirror symmetry for log Calabi–Yau surfaces I. *Publ. Math. Inst. Hautes Études Sci.*, 122:65–168, 2015.

[49] V. Guillemin and S. Sternberg. Convexity properties of the moment mapping. *Invent. Math.*, 67(3):491–513, 1982.

[50] V. Guillemin and S. Sternberg. Birational equivalence in the symplectic category. *Invent. Math.*, 97(3):485–522, 1989.

[51] P. Hacking and Y. Prokhorov. Smoothable del Pezzo surfaces with quotient singularities. *Compos. Math.*, 146(1):169–92, 2010.

[52] P. Hacking, J. Tevelev, and G. Urzúa. Flipping surfaces. *J. Algebraic Geom.*, 26(2):279–345, 2017.

[53] P. R. Halmos. *Finite dimensional vector spaces*, volume 7 of Annals of Mathematics Studies. Princeton University Press, Princeton, NJ, 1942.

[54] F. E. P. Hirzebruch. Hilbert modular surfaces. *Enseign. Math. (2)*, 19(3–4):183–281, 1973.

[55] M. Jankins and W. D. Neumann. *Lectures on Seifert manifolds*, volume 2 of Brandeis Lecture Notes. Brandeis University, Waltham, MA, 1983.

[56] D. Joyce. Singularities of special Lagrangian fibrations and the SYZ conjecture. *Comm. Anal. Geom.*, 11(5):859–907, 2003.

[57] D. Karabas. Microlocal sheaves on pinwheels. *arXiv:1810.09021*, 2018.

[58] T. Khodorovskiy. Bounds on embeddings of rational homology balls in symplectic 4-manifolds. *arXiv:1307.4321*, 2013.

[59] J. Kollár and N. I. Shepherd-Barron. Threefolds and deformations of surface singularities. *Invent. Math.*, 91(2):299–338, 1988.

[60] M. Konstantinov. Symplectic Topology of Projective Space: Lagrangians, Local Systems and Twistors. PhD thesis, University College London, 2019.

[61] M. Kontsevich and Y. Soibelman. Homological mirror symmetry and torus fibrations. In K. Fukaya, Y.-G. Oh, K. Ono, and G. Tian, editors, *Symplectic geometry and mirror symmetry (Seoul, 2000)*, pages 203–63. World Scientific, River Edge, NJ, 2001.

[62] R. Lazarsfeld. *Positivity in algebraic geometry. I*, volume 48 of Ergebnisse der Mathematik und ihrer Grenzgebiete. 3. Folge. Springer, Berlin, 2004.

[63] J. M. Lee. *Introduction to smooth manifolds*, volume 218 of Graduate Texts in Mathematics. Springer, New York, 2nd ed., 2013.

[64] Y. Lekili and M. Maydanskiy. The symplectic topology of some rational homology balls. *Comment. Math. Helv.*, 89(3):571–96, 2014.

[65] E. Lerman. Symplectic cuts. *Math. Res. Lett.*, 2(3):247–58, 1995.

[66] N. C. Leung and M. Symington. Almost toric symplectic four-manifolds. *J. Symplectic Geom.*, 8(2):143–87, 2010.

[67] T.-J. Li and Y. Ruan. Symplectic birational geometry. In M. Abreu, F. Lalonde, and L. Polterovich, editors, *New perspectives and challenges in symplectic field theory*, volume 49 of CRM Proceedings Lecture Notes, pages 307–26. American Mathematical Society, Providence, RI, 2009.

[68] P. Lisca. On symplectic fillings of lens spaces. *Trans. Amer. Math. Soc.*, 360(2):765–99, 2008.

[69] E. Looijenga and J. Wahl. Quadratic functions and smoothing surface singularities. *Topology*, 25(3):261–91, 1986.

[70] N. Magill and D. McDuff. Staircase symmetries in Hirzebruch surfaces. *arXiv:2106.09143*, 2021.

[71] M. Manetti. Normal degenerations of the complex projective plane. *J. Reine Angew. Math.*, 419:89–118, 1991.

[72] D. Matessi. Lagrangian pairs of pants. *Int. Math. Res. Not. IMRN*, 2021(15):11306–56.

[73] D. Matessi. Lagrangian submanifolds from tropical hypersurfaces. *Internat. J. Math.*, 32(7): Paper No. 2150046, 2021.

[74] D. McDuff. The structure of rational and ruled symplectic 4-manifolds. *J. Amer. Math. Soc.*, 3(3):679–712, 1990.

[75] D. McDuff and L. Polterovich. Symplectic packings and algebraic geometry. *Invent. Math.*, 115(3):405–34, 1994. With an appendix by Yael Karshon.

[76] D. McDuff and D. Salamon. *J-holomorphic curves and quantum cohomology*, volume 6 of University Lecture Series. American Mathematical Society, Providence, RI, 1994.

[77] D. McDuff and D. Salamon. *Introduction to symplectic topology.*. Oxford University Press, New York, 2nd ed., 1998.

[78] D. McDuff and F. Schlenk. The embedding capacity of 4-dimensional symplectic ellipsoids. *Ann. of Math. (2)*, 175(3):1191–282, 2012.

[79] M. McLean. Birational Calabi–Yau manifolds have the same small quantum products. *Ann. of Math. (2)*, 191(2):439–579, 2020.

[80] G. Mikhalkin. Examples of tropical-to-Lagrangian correspondence. *Eur. J. Math.*, 5(3):1033–66, 2019.

[81] J. Milnor. *Singular points of complex hypersurfaces*, volume 61 of Annals of Mathematics Studies. Princeton University Press, Princeton, NJ; University of Tokyo Press, Tokyo, 1968.

[82] J. Moser. On the volume elements on a manifold. *Trans. Amer. Math. Soc.*, 120:286–94, 1965.

[83] P. Orlik. *Seifert manifolds*, volume 291 of Lecture Notes in Mathematics. Springer, Berlin, 1972.

[84] J. Park. Simply connected symplectic 4-manifolds with $b_2^+ = 1$ and $c_1^2 = 2$. *Invent. Math.*, 159(3):657–67, 2005.

[85] J. Pascaleff and D. Tonkonog. The wall-crossing formula and Lagrangian mutations. *Adv. Math.*, 361:106850, 2020.

[86] J. Rana and G. Urzúa. Optimal bounds for T-singularities in stable surfaces. *Adv. Math.*, 345:814–44, 2019.

[87] W.-D. Ruan. Lagrangian torus fibration of quintic hypersurfaces. I. Fermat quintic case. In C. Vafa and S.-T. Yau, editors, *Winter School on Mirror Symmetry, Vector Bundles and Lagrangian Submanifolds (Cambridge, MA, 1999)*, volume 23 of AMS/IP Studies in Advanced Mathematics, pages 297–332. American Mathematical Society, Providence, RI, 2001.

[88] W.-D. Ruan. Lagrangian torus fibration of quintic Calabi–Yau hypersurfaces. II. Technical results on gradient flow construction. *J. Symplectic Geom.*, 1(3):435–521, 2002.

[89] W.-D. Ruan. Lagrangian torus fibration of quintic Calabi–Yau hypersurfaces. III. Symplectic topological SYZ mirror construction for general quintics. *J. Differential Geom.*, 63(2):171–229, 2003.

[90] A. N. Rudakov. Markov numbers and exceptional bundles on $\mathbf{P}^2$. *Izv. Akad. Nauk SSSR Ser. Mat.*, 52(1):100–12, 240, 1988.

[91] F. Schlenk. *Embedding problems in symplectic geometry*, volume 40 of De Gruyter Expositions in Mathematics. Walter de Gruyter GmbH & Co. KG, Berlin, 2005.

[92] R. Schoen and J. Wolfson. Minimizing area among Lagrangian surfaces: The mapping problem. *J. Differential Geom.*, 58(1):1–86, 2001.

[93] P. Seidel. Disjoinable Lagrangian spheres and dilations. *Invent. Math.*, 197(2):299–359, 2014.

[94] H. Seifert and W. Threlfall. *Seifert and Threlfall: a textbook of topology*, volume 89 of Pure and Applied Mathematics. Academic Press, Inc. [Harcourt Brace Jovanovich, Publishers], New York-London, 1980. Translated from the German edition of 1934 by Michael A. Goldman, with a preface by Joan S. Birman, With 'Topology of 3-dimensional fibered spaces' by Seifert, translated from the German by Wolfgang Heil.

[95] D. Sepe and S. Vũ Ngọc. Integrable systems, symmetries, and quantization. *Lett. Math. Phys.*, 108(3):499–571, 2018.

[96] C. Series. The geometry of Markoff numbers. *Math. Intelligencer*, 7(3):20–9, 1985.

[97] V. Shende, D. Treumann, and H. Williams. On the combinatorics of exact Lagrangian surfaces. *arXiv:1603.07449*, 2016.

[98] V. Shevchishin and G. Smirnov. Symplectic triangle inequality. *Proc. Amer. Math. Soc.*, 148(4):1389–97, 2020.

[99] V. V. Shevchishin. On the local Severi problem. *Int. Math. Res. Not.*, 2004(5):211–37, 2004.

[100] B. Siebert and G. Tian. On the holomorphicity of genus two Lefschetz fibrations. *Ann. of Math. (2)*, 161(2):959–1020, 2005.

[101] B. Siebert and G. Tian. Lectures on pseudo-holomorphic curves and the symplectic isotopy problem. In *Symplectic 4-manifolds and algebraic surfaces*, volume 1938 of Lecture Notes in Mathematics, pages 269–341. Springer, Berlin, 2008.

[102] J.-C. Sikorav. The gluing construction for normally generic J-holomorphic curves. In Y. Eliashberg, B. Khesin, and F. Lalonde, editors, *Symplectic and contact topology: Interactions and perspectives (Toronto, ON/Montreal, QC, 2001)*, volume 35 of Fields Institute Communications, pages 175–99. American Mathematical Society, Providence, RI, 2003.

[103] J. Smith. Floer cohomology of Platonic Lagrangians. *J. Symplectic Geom.*, 17(2):477–601, 2019.

[104] J. Stevens. The versal deformation of cyclic quotient singularities. In A. Némethi and Á. Szilárd, editors, *Deformations of surface singularities*, volume 23 of Bolyai Society Mathematical Studies, pages 163–201. János Bolyai Mathematical Society, Budapest, 2013.

[105] M. Symington. Generalized symplectic rational blowdowns. *Algebr. Geom. Topol.*, 1:503–18, 2001.

[106] M. Symington. Four dimensions from two in symplectic topology. In G. Matić and C. McCrory, editors, *Topology and geometry of manifolds (Athens, GA, 2001)*, volume 71 of Proceedings of Symposia in Pure Mathematics, pages 153–208. American Mathematical Society, Providence, RI, 2003.

[107] W. P. Thurston. Some simple examples of symplectic manifolds. *Proc. Amer. Math. Soc.*, 55(2):467–8, 1976.

[108] G. Urzúa. Identifying neighbors of stable surfaces. *Ann. Sc. Norm. Super. Pisa Cl. Sci. (5)*, 16(4):1093–122, 2016.

[109] G. Urzúa and N. Vilches. On wormholes in the moduli space of surfaces. *Algebr. Geom.*, 9(1):39–68, 2022.

[110] M. Usher. Symplectic blow-up. MathOverflow. https://mathoverflow.net/q/133709 (version: 2013-06-14).

[111] S. Vũ Ngọc. On semi-global invariants for focus–focus singularities. *Topology*, 42(2):365–80, 2003.

[112] San Vũ Ngọc. Bohr-Sommerfeld conditions for integrable systems with critical manifolds of focus–focus type. *Comm. Pure Appl. Math.*, 53(2):143–217, 2000.

[113] U. Varolgunes. Mayer-Vietoris property for relative symplectic cohomology. *Geom. Topol.*, 25(2):547–642, 2021.

[114] S. Venkatesh. Rabinowitz Floer homology and mirror symmetry. *J. Topol.*, 11(1):144–79, 2018.

[115] R. Vianna. On exotic Lagrangian tori in $\mathbb{CP}^2$. *Geom. Topol.*, 18(4):2419–76, 2014.

[116] R. Vianna. Infinitely many exotic monotone Lagrangian tori in $\mathbb{CP}^2$. *J. Topol.*, 9(2):535–51, 2016.

[117] R. Vianna. Infinitely many monotone Lagrangian tori in del Pezzo surfaces. *Sel. Math., New Ser.*, 23(3):1955–96, 2017.

[118] F. W. Warner. *Foundations of differentiable manifolds and Lie groups*, volume 94 of Graduate Texts in Mathematics. Springer, Berlin, 1983. Corrected reprint of the 1971 edition.

[119] E. T. Whittaker. *A treatise on the analytical dynamics of particles and rigid bodies: With an introduction to the problem of three bodies.* Cambridge University Press, New York, 1959. 4th ed.

[120] N. T. Zung. Symplectic topology of integrable Hamiltonian systems. I. Arnold-Liouville with singularities. *Compositio Math.*, 101(2):179–215, 1996.

[121] N. T. Zung. Symplectic topology of integrable Hamiltonian systems. II. Topological classification. *Compositio Math.*, 138(2):125–56, 2003.

Index

221

Printed in the United States
by Baker & Taylor Publisher Services

Printed in the United States
by Baker & Taylor Publisher Services